The Child and Television Drama: The Psychosocial Impact of Cumulative Viewing

Formulated by the Committee on Social Issues
Group for the Advancement of Psychiatry

Mental Health Materials Center
30 E. 29th Street, New York, N.Y. 10016

Library of Congress Cataloging in Publication Data

Group for the Advancement of Psychiatry. Committee
  on Social Issues.
  The child and television drama.

  (Publication; v. 11, no. 112)
  Bibliography: p.
  1. Television and children--United States--Psychological aspects. 2. Television and children--Social aspects--United States. 3. Child development. I. Title.
II. Series: Publication (Group for the Advancement of Psychiatry); no. 112. [DNLM: 1. Television. 2. Child development. 3. Child behavior. W1 PU675N
no. 112/WS 105.5.E9 C536]
HQ784.T4G76   1982   305.2'3   82-21717
ISBN 0-910958-17-3

Copyright © 1982 by the Mental Health Materials Center. All rights reserved. This report must not be reproduced in any form without written permission of the Mental Health Materials Center, except by a reviewer, reporter, or commentator who wishes to quote brief passages.

November, 1982, Volume XI, Publication No. 112

This is the sixth in a series of GAP publications comprising Volume XI.

Manufactured in the United States of America.

TABLE OF CONTENTS

STATEMENT OF PURPOSE vi

COMMITTEE ACKNOWLEDGMENTS xii

FOREWORD 1

1 THE MEDIUM: TELEVISION IN AMERICA 5
    Capacity and power of the medium 5
    The uniqueness of television 7
    The economics of the television industry 8
    Television as a psychosocial issue 12

2 THE AUDIENCE: CHILDREN AS VIEWERS 17
    Needs of the developing child 18
        The changing family environment 19
        The psychological process of identification 20
    Stages of psychological development 23
        Infancy and toddlerhood 24
        Preschool Years 25
        School age 26
        Adolescence 29
        Summary 32

3 THE AUDIENCE: SOME QUESTIONS OF EFFECT 35
    How does television affect play and fantasy? 35
        Why fantasize? 35
        Images of fantasy and reality 36
        When fiction and reality merge 37
        The ability to fantasize 39
    Does television promote passivity? 42
    Does television foster dependent personalities? 44
    How does television relate to loneliness? 47

4 THE PROGRAMS: DRAMAS WITH LIVE ACTORS . 52
   The plots 53
      Conflict and conflict resolution 53
         Conflict resolution 55
      The role of violence 58
         Prevalence and significance 58
         Effects of viewer aggression 60
         Research findings 62
         Some other effects 66
         Why violence? 68
   The characters 70
      Stereotypes 71
         The work force 71
         Men and women 72
         Minorities 73
         Feelings and traits 74
         Dehumanization 76

5 THE PROGRAMS: CHILDREN'S TELEVISION 83
   Cartoons 83
   Educational programs 90
      How can television educate? 90
         Building knowledge and skills 90
         Promoting healthy psychological development . . . 90
         Fostering positive social relations 92
      Children's educational programs 94
         Some concerns 96
      A programming need: promoting mental health . . . 99

6 RECOMMENDATIONS AND GUIDELINES 105
   Recommendations and guidelines for parents 106
      Some guidelines 109
         1. Assess the role of television in the child's life . . 109
         2. Evaluate what the child is watching 110
         3. Set a limit on the amount of viewing 111

Contents

    4. Monitor and share in the child's viewing
       experiences 111
Recommendations to mental health professionals 114
Recommendations to public policymakers 115
Recommendations to the television industry 116
A final note 118

APPENDIX 120

STATEMENT OF PURPOSE

THE GROUP FOR THE ADVANCEMENT OF PSYCHIATRY has a membership of approximately 300 psychiatrists, most of whom are organized in the form of a number of working committees. These committees direct their efforts toward the study of various aspects of psychiatry and the application of this knowledge to the fields of mental health and human relations.

Collaboration with specialists in other disciplines has been and is one of GAP's working principles. Since the formation of GAP in 1946 its members have worked closely with such other specialists as anthropologists, biologists, economists, statisticians, educators, lawyers, nurses, psychologists, sociologists, social workers, and experts in mass communication, philosophy, and semantics. GAP envisages a continuing program of work according to the following aims:

1. To collect and appraise significant data in the fields of psychiatry, mental health, and human relations
2. To reevaluate old concepts and to develop and test new ones
3. To apply the knowledge thus obtained for the promotion of mental health and good human relations.

GAP is an independent group, and it reports represent the composite findings and opinions of its members only, guided by its many consultants.

THE CHILD AND TELEVISION DRAMA: THE PSYCHOSOCIAL IMPACT OF CUMULATIVE VIEWING was formulated by the Committee on Social Issues, which acknowledges on page xii the participation of others in the preparation of this report. The

Statement of Purpose vii

members of this committee are listed below. The following pages list the members of the other GAP committees as well as additional membership categories and current and past officers of GAP.

COMMITTEE ON SOCIAL ISSUES

Roy W. Menninger, Topeka, Kans.,
   Chairperson
Ian A. Alger, New York, N.Y.
William R. Beardslee, Boston, Mass.
Viola W. Bernard, New York, N.Y.
Paul Jay Fink, Philadelphia, Pa.
Henry J. Gault, Chicago, Ill.
Roderick Gorney, Los Angeles, Calif.
Lester Grinspoon, Boston, Mass.
Joel Handler, Chicago, Ill.
Judd Marmor, Los Angeles, Calif.
Perry Ottenberg, Philadelphia, Pa.
Kendon Smith, New York, N.Y.

COMMITTEE ON ADOLESCENCE

Warren J. Gadpaille, Englewood, Colo.,
   Chairperson
Ian A. Canino, New York, N.Y.
Harrison P. Eddy, New York, N.Y.
Sherman C. Feinstein, Highland Park, Ill.
Michael Kalogerakis, New York, N.Y.
Clarice J. Kestenbaum, New York, N.Y.
Derek Miller, Chicago, Ill.
Silvio J. Onesti, Jr., Belmont, Mass.

COMMITTEE ON AGING

Charles M. Gaitz, Houston, Tex., Chairperson
Gene D. Cohen, Rockville, Md.
Lawrence F. Greenleigh, Los Angeles, Calif.
George H. Pollock, Chicago, Ill.
Harvey L. Ruben, New Haven, Conn.
F. Conyers Thompson, Jr., Atlanta, Ga.

COMMITTEE ON CHILD PSYCHIATRY

John F. McDermott, Jr., Honolulu, Hawaii,
   Chairperson
Paul L. Adams, Louisville, Ky.
James M. Bell, Canaan, N.Y.
Harlow Donald Dunton, New York, N.Y.
Joseph Fischoff, Detroit, Mich.
Joseph M. Green, Madison, Wis.
John Schowalter, New Haven, Conn.
Theodore Shapiro, New York, N.Y.
Peter Tanguay, Los Angeles, Calif.
Lenore F. C. Terr, San Francisco, Calif.

COMMITTEE ON THE COLLEGE STUDENT

Kent E. Robinson, Towson, Md.,
   Chairperson
Robert L. Arnstein, Hamden, Conn.
Varda Backus, La Jolla, Calif.
Myron B. Liptzin, Chapel Hill, N.C.
Malkah Tolpin Notman, Brookline, Mass.
Gloria C. Onque, Pittsburgh, Pa.
Elizabeth Aub Reid, Cambridge, Mass.
Earle Silber, Chevy Chase, Md.

COMMITTEE ON CULTURAL PSYCHIATRY

Andrea K. Delgado, New York, N.Y.,
   Chairperson
John P. Spiegel, Waltham, Mass.
Ronald M. Wintrob, Farmington, Conn.

COMMITTEE ON THE FAMILY

Henry U. Grunebaum, Cambridge, Mass.,
   Chairperson
W. Robert Beavers, Dallas, Tex.
Ellen M. Berman, Merion, Pa.
Lee Combrinck-Graham, Philadelphia, Pa.
Ira D. Glick, New York, N.Y.
Frederick Gottlieb, Los Angeles, Calif.
Charles A. Malone, Cleveland, Ohio
Joseph Satten, San Francisco, Calif.

COMMITTEE ON GOVERNMENTAL AGENCIES

William W. Van Stone, Palo Alto, Calif., Chairperson,
James P. Cattell, Monterey, Mass.
Sidney S. Goldensohn, Jamaica, N.Y.
Naomi Heller, Washington, D.C.
Roger Peele, Washington, D.C.

COMMITTEE ON HANDICAPS

Norman R. Bernstein, Chicago, Ill., Chairperson
Betty J. Pfefferbaum, Houston, Tex.
George Tarjan, Los Angeles, Calif.
Thomas G. Webster, Washington, D.C.
Henry H. Work, Washington, D.C.

COMMITTEE ON INTERNATIONAL RELATIONS

Francis F. Barnes, Chevy Chase, Md., Chairperson
Robert M. Dorn, Norfolk, Va.
John E. Mack, Chestnut Hill, Mass.
Rita R. Rogers, Torrance, Calif.
Bertram H. Schaffner, New York, N.Y.
Stephen B. Shanfield, Tucson, Ariz.
Vamik D. Volkan, Charlottesville, Va.

COMMITTEE ON MEDICAL EDUCATION

David R. Hawkins, Chicago, Ill., Chairperson
Carol Nadelson, Boston, Mass.
Herbert Pardes, Rockville, Md.
Carolyn Robinowitz, Bethesda, Md.
Sidney L. Werkman, Denver, Colo.

COMMITTEE ON MENTAL HEALTH SERVICES

Herzl R. Spiro, Milwaukee, Wis., Chairperson
Mary Ann B. Bartusis, Morrisville, Pa.
Allan Beigel, Tucson, Ariz.
Eugene M. Caffey, Jr., Bowie, Md.
Merrill T. Eaton, Omaha, Nebr.
Joseph T. English, New York, N.Y.
W. Walter Menninger, Topeka, Kans.
Jose Maria Santiago, Tucson, Ariz.
George F. Wilson, Belle Mead, N.J.
Jack A. Wolford, Pittsburgh, Pa.

COMMITTEE ON PREVENTIVE PSYCHIATRY

Stephen Fleck, New Haven, Conn., Chairperson
C. Knight Aldrich, Charlottesville, Va.
Viola W. Bernard, New York, N.Y.
Jules V. Coleman, New Haven, Conn.
William H. Hetznecker, Philadelphia, Pa.
Richard G. Morrill, Boston, Mass.
Harris B. Peck, New Rochelle, N.Y.

COMMITTEE ON PSYCHIATRY AND COMMUNITY

John A. Talbott, New York, N.Y., Chairperson
John C. Nemiah, Boston, Mass.
Alexander S. Rogawski, Los Angeles, Calif.
John J. Schwab, Louisville, Ky.
Charles B. Wilkinson, Kansas City, Mo.

COMMITTEE ON PSYCHIATRY AND LAW

Loren H. Roth, Pittsburgh, Pa., Chairperson
Elissa P. Benedek, Ann Arbor, Mich.
Park E. Dietz, Belmont, Mass.
John Donnelly, Hartford, Conn.
Seymour L. Halleck, Chapel Hill, N.C.
Carl P. Malmquist, Minneapolis, Minn.
A. Louis McGarry, Great Neck, N.Y.
Herbert C. Modlin, Topeka, Kans.
Jonas R. Rappeport, Baltimore, Md.
William D. Weitzel, Lexington, Ky.

COMMITTEE ON PSYCHIATRY AND RELIGION

Albert J. Lubin, Woodside, Calif., Chairperson
Sidney Furst, Bronx, N.Y.
Richard C. Lewis, New Haven, Conn.
Mortimer Ostow, Bronx, N.Y.
Clyde R. Snyder, Columbus, Mo.
Michael R. Zales, Greenwich, Conn.

COMMITTEE ON PSYCHIATRY IN INDUSTRY

Duane Q. Hagen, St. Louis, Mo., Chairperson
Barrie S. Greiff, Boston, Mass.
R. Edward Huffman, Asheville, N.C.
Herbert L. Klemme, Stillwater, Minn.
Alan A. McLean, New York, N.Y.
David E. Morrison, Palatine, Ill.
Clarence J. Rowe, St. Paul, Minn.
John Wakefield, Saratoga, Calif.

Statement of Purpose

COMMITTEE ON PSYCHOPATHOLOGY

David A. Adler, Boston, Mass., Chairperson
Wagner H. Bridger, Bronx, N.Y.
Doyle I. Carson, Dallas, Tex.
Paul E. Huston, Iowa City, Iowa
Richard E. Renneker, Los Angeles, Calif.

COMMITTEE ON PUBLIC EDUCATION

Robert J. Campbell, New York, N.Y., Chairperson
Norman L. Loux, Sellersville, Pa.
Julius Schreiber, Washington, D.C.
Miles F. Shore, Boston, Mass.
Robert A. Solow, Beverly Hills, Calif.
Kent A. Zimmerman, Berkeley, Calif.

COMMITTEE ON RESEARCH

Jerry M. Lewis, Dallas, Tex., Chairperson
John E. Adams, Gainesville, Fla.
Robert Cancro, New York, N.Y.
Stanley H. Eldred, Belmont, Mass.
John G. Gunderson, Belmont, Mass.
Morris A. Lipton, Chapel Hill, N.C.
John G. Looney, Dallas, Tex.
Charles P. O'Brien, Philadelphia, Pa.
Alfred H. Stanton, Wellesley Hills, Mass.
John S. Strauss, New Haven, Conn.

COMMITTEE ON THERAPEUTIC CARE

Orlando B. Lightfoot, Boston, Mass., Chairperson
Bernard Bandler, Cambridge, Mass.
Thomas E. Curtis, Chapel Hill, N.C.
Robert W. Gibson, Towson, Md.
Donald W. Hammersley, Washington, D.C.
Roberto L. Jimenez, San Antonio, Tex.
Milton Kramer, Cincinnati, Ohio
Melvin Sabshin, Washington, D.C.

COMMITTEE ON THERAPY

Robert Michels, New York, N.Y., Chairperson
Henry W. Brosin, Tucson, Ariz.
James S. Eaton, Jr., Rockville, Md.
Eugene B. Feigelson, Brooklyn, N.Y.
Tokoz Byram Karasu, New York, N.Y.
Andrew P. Morrison, Cambridge, Mass.
William C. Offenkrantz, Milwaukee, Wis.
Lewis L. Robbins, Glen Oaks, N.Y.
Allan D. Rosenblatt, La Jolla, Calif.

CONTRIBUTING MEMBERS

Carlos C. Alden, Jr., Buffalo, N.Y.
Charlotte G. Babcock, Pittsburgh, Pa.
Grace Baker, New York, N.Y.
Eric A. Baum, Akron, Ohio
Spencer Bayles, Houston, Tex.
Aaron T. Beck, Wynnewood, Pa.
C. Christian Beels, New York, N.Y.
Sidney Berman, Washington, D.C.
Wilfred Bloomberg, Cambridge, Mass.
Thomas L. Brannick, Imola, Calif.
H. Keith H. Brodie, Durham, N.C.
C. Martel Bryant, San Francisco, Calif.
Ewald W. Busse, Durham, N.C.
Robert N. Butler, Washington, D.C.
Paul Chodoff, Washington, D.C.
Ian L. W. Clancy, Ontario, Canada
Sanford I. Cohen, Boston, Mass.
William D. Davidson, Washington, D.C.
Lloyd C. Elam, Nashville, Tenn.
Louis C. English, Pomona, N.Y.
Raymond Feldman, Boulder, Colo.
Alfred Flarsheim, Wilmette, Ill.
Archie R. Foley, New York, N.Y.
Alan Frank, Albuquerque, N.M.
Daniel X. Freedman, Chicago, Ill.
James B. Funkhouser, Richmond, Va.
Robert S. Garber, Belle Mead, N.J.
Albert J. Glass, Bethesda, Md.
Alexander Gralnick, Port Chester, N.Y.
Milton Greenblatt, Los Angeles, Calif.

Statement of Purpose

John H. Greist, Indianapolis, Ind.
Roy R. Grinker, Sr., Chicago, Ill.
Lester Grinspoon, Boston, Mass.
Ernest M. Gruenberg, Baltimore, Md.
Stanley Hammons, Lexington, Ky.
Joel S. Handler, Wilmette, Ill.
Saul I. Harrison, Ann Arbor, Mich.
Peter Hartocollis, Topeka, Kans.
J. Cotter Hirschberg, Topeka, Kans.
Edward J. Hornick, New York, N.Y.
Joseph Hughes, Philadelphia, Pa.
Portia Bell Hume, St. Helena, Calif.
*Benjamin Jeffries, Harper Woods, Mich.
Jay Katz, New Haven, Conn.
Sheppard G. Kellam, Chicago, Ill.
Peter H. Knapp, Boston, Mass.
James A. Knight, New Orleans, La.
Othilda M. Krug, Cincinnati, Ohio
Robert L. Leopold, Philadelphia, Pa.
Alan I. Levenson, Tucson, Ariz.
Ruth W. Lidz, Woodbridge, Conn.
Maurice E. Linden, Philadelphia, Pa.
Earl A. Loomis, Jr., Greenport, N.Y.
Reginald S. Lourie, Chevy Chase, Md.
Jeptha R. MacFarlane, Garden City, N.Y.
John A. MacLeod, Cincinnati, Ohio
Leo Madow, Philadelphia, Pa.
Sidney G. Margolin, Denver, Colo.
Peter A. Martin, Southfield, Mich.
Ake Mattsson, New York, N.Y.
David Mendell, Houston, Tex.
Mary E. Mercer, Nyack, N.Y.
*Eugene Meyer, Baltimore, Md.
James G. Miller, Louisville, Ky.
John E. Nardini, Washington, D.C.
Joseph D. Noshpitz, Washington, D.C.
Lucy D. Ozarin, Bethesda, Md.
Bernard L. Pacella, New York, N.Y.
Norman L. Paul, Boston, Mass.
Marvin E. Perkins, Roanoke, Va.
Charles A. Pinderhughes, Bedford, Mass.
Seymour Pollack, Los Angeles, Calif.
David N. Ratnavale, Chevy Chase, Md.
Walter Reich, Rockville, Md.
Harvey L. P. Resnik, College Park, Md.
W. Donald Ross, Cincinnati, Ohio
Lester H. Rudy, Chicago, Ill.
George E. Ruff, Philadelphia, Pa.

*Deceased

A. John Rush, Dallas, Tex.
David S. Sanders, Los Angeles, Calif.
Donald Scherl, Brooklyn, N.Y.
Kurt O. Schlesinger, San Francisco, Calif.
*Robert A. Senescu, Albuquerque, N. Mex.
Calvin F. Settlage, Sausalito, Calif.
Charles Shagass, Philadelphia, Pa.
Albert J. Silverman, Ann Arbor, Mich.
Justin Simon, Berkeley, Calif.
Kendon W. Smith, Valhalla, N.Y.
Benson R. Snyder, Cambridge, Mass.
David A. Soskis, Bala Cynwyd, Pa.
Jeanne Spurlock, Washington, D.C.
Tom G. Stauffer, White Plains, N.Y.
Brandt F. Steele, Denver, Colo.
Eleanor A. Steele, Denver, Colo.
Rutherford B. Stevens, New York, N.Y.
Alan A. Stone, Cambridge, Mass.
Robert E. Switzer, Trevose, Pa.
Perry C. Talkington, Dallas, Tex.
Graham C. Taylor, Montreal, Canada
Prescott W. Thompson, Beaverton, Oreg.
Harvey J. Tompkins, New York, N.Y.
Lucia E. Tower, Chicago, Ill.
Joseph P. Tupin, Sacramento, Calif.
John A. Turner, San Francisco, Calif.
Montague Ullman, Ardsley, N.Y.
Gene L. Usdin, New Orleans, La.
Warren T. Vaughan, Jr., Portola Valley, Calif.
Robert S. Wallerstein, San Francisco, Calif.
Andrew S. Watson, Ann Arbor, Mich.
Bryant M. Wedge, Washington, D.C.
Joseph B. Wheelwright, Kentfield, Calif.
Robert L. Williams, Houston, Tex.
Paul Tyler Wilson, Bethesda, Md.
Sherwyn M. Woods, Los Angeles, Calif.
Stanley F. Yolles, Stony Brook, N.Y.
Israel Zwerling, Philadelphia, Pa.

LIFE MEMBERS

*S. Spafford Ackerly, Louisville, Ky.
C. Knight Aldrich, Charlottesville, Va.
Bernard Bandler, Cambridge, Mass.
Leo H. Bartemeier, Baltimore, Md.
Walter E. Barton, Hartland, Vt.
Ivan C. Berlien, Coral Gables, Fla.
Murray Bowen, Chevy Chase, Md.
O. Spurgeon English, Narberth, Pa.

Statement of Purpose

Dana L. Farnsworth, Belmont, MA
Stephen Fleck, New Haven, Conn.
Jerome Frank, Baltimore, Md.
Edward O. Harper, Cleveland, Ohio
Margaret M. Lawrence, Pomona, N.Y.
Harold I. Lief, Philadelphia, Pa.
Judd Marmor, Los Angeles, Calif.
Karl A. Menninger, Topeka, Kans.
Lewis L. Robbins, Glen Oaks, N.Y.
Mabel Ross, Sun City, Ariz.
Francis A. Sleeper, Cape Elizabeth, Maine

LIFE CONSULTANT

*Mrs. Ethel L. Ginsburg, New York, N.Y.

BOARD OF DIRECTORS

Officers

President

Henry H. Work
1700 Eighteenth Street, N.W.
Washington, D.C. 20009

Vice-President

Robert W. Gibson
Sheppard and Enoch Pratt Hospital
Towson, Md. 21204

Secretary

Allan Beigel
30 Camino Espanole
Tucson, Ariz. 85716

Treasurer

Michael R. Zales
Edgewood Drive
Greenwich, Conn. 06830

Immediate Past President, 1981–82

*Jack Weinberg, 1981–82

*Deceased

Board Members

Donald W. Hammersley
Malkah Tolpin Notman
William C. Offenkrantz
Ronald M. Wintrob

Past Presidents

*William C. Menninger	1946–51
Jack R. Ewalt	1951–53
Walter E. Barton	1953–55
*Sol W. Ginsburg	1955–57
Dana L. Farnsworth	1957–59
*Marion E. Kenworthy	1959–61
Henry W. Brosin	1961–63
Leo H. Bartemeier	1963–65
Robert S. Garber	1965–67
Herbert C. Modlin	1967–69
John Donnelly	1969–71
George Tarjan	1971–73
Judd Marmor	1973–75
John C. Nemiah	1975–77
Jack A. Wolford	1977–79
Robert W. Gibson	1979–81

PUBLICATIONS BOARD

Chairman

Merrill T. Eaton
Nebraska Psychiatric Institute
602 South 45th Street
Omaha, Neb. 68106

C. Knight Aldrich
Robert Arnstein
Carl P. Malmquist
Perry Ottenberg
Alexander S. Rogawski

Consultant

John C. Nemiah

Ex-Officio

Robert W. Gibson
Henry H. Work

COMMITTEE ACKNOWLEDGMENTS

For their counsel and participation in the formulation of this report, we wish to acknowledge with deep gratitude the contributions of Fritz Redl, Ph.D., our long time consultant, and our deceased colleague, Arthur A. Miller, M.D. We also wish to acknowledge the participation of our Ginsburg Fellows, Alvin F. Heap, M.D., and Miklos Losonczy, M.D. in the deliberations which led to the formulation of this work. Our special thanks are due to Carolyn Mercer-McFadden, Ph.D., for the excellent editorial assistance which she provided in preparing the final version of this report.

Roy W. Menninger, Chairperson

FOREWORD

Our species is imperiled by its own actions. Overpopulation, pollution, the exhaustion of the earth's natural resources and the threat of nuclear war are increasingly evident dangers. Although crisis often stimulates change, it would be folly to assume that a crisis great enough to awaken us to the need for change will not also destroy us. The perils we face are unprecedented in scope and complexity. If we are to avoid self-destruction, we must mobilize our ability to foresee, to understand, to plan, and to act.

H.G. Wells noted that "human history becomes more and more a race between education and catastrophe."[1] Although the possibility that education could create the conditions for human survival has a long tradition and a large constituency, its process has been too slow to overtake the rate at which dangers multiply. Television presents an opportunity to accelerate the pace and improve the impact of the education upon which survival depends. It offers a means for contemporary education that is more effective and extensive than any yet envisioned. It is unparalleled in its capacity to present imaginative examples of alternative futures, to model effective interpersonal relationships, and to depict critical social choices and trade-offs.

But this is an image of what television could be and perhaps someday will be. It is not what it is today. The contemporary picture is both less optimistic and more troubling. Television programming is overwhelmingly dominated by commercially profitable presentations intended to entertain. It does not demonstrate much use of the positive potential of television for education. Nor are there prominent trends toward qualitative improvement. Achieving this potential will require more adequately financed and more relevant research and development than has yet been undertaken, and a collective commitment, not yet evident, to move toward such an objective.

As physicians continually contending with psychopathology, psychiatrists are particularly sensitive to the pathogenic possibilities of television programs. But psychiatrists also recognize that television has a positive potential—a potential which can enhance health and promote improvement in the quality of life.

This report, written by psychiatrists, is our contribution to the growing body of knowledge about television—a viewpoint on an aspect of special concern to us: the psychosocial effects of cumulative viewing. We focus on one important type of television programming, drama, as it affects a particularly important American audience, children.

Reporting in 1969, the National Commission on the Causes and Prevention of Violence concluded, "In a fundamental way, television helps to create what children expect of themselves and of others and what constitutes the standards of civilized society."[2] We concur with this judgment. Likewise, television in the long run helps shape the actual characteristics of society itself, civilized or otherwise.

As psychiatrists, our interest in the effects of television viewing on children also stems from our concern for their well-being as individuals. We are troubled by the potentially harmful effects of so much of the programming itself. And we are troubled that so many children spend so much time watching television, to the possible detriment of their balanced development.

We have special concern about the dramatic programs that children watch, although we nourish some hope about them too. Because the dramas evoke strong identifications and intense emotional responses in the viewer, they have great possibilities for facilitating human growth, as well as for doing damage.

The tragic dramas of Sophocles, the comedies of Aristophanes, and both the tragedies and comedies of Shakespeare go far beyond entertainment. While it may be unrealistic to hold a medium designed to sell products to consumers to such high standards, for perspective it is well to bear in mind the dimensions of the classical tradition. For centuries dramas have brought to their audiences sharply drawn questions about existence, values, and meaning. They have defined and portrayed both the frailties of human nature and its capacity for triumph. Classical dramas have offered a means for humankind to relate itself to the larger universe. We believe that television drama could do the same.

Foreword

We are hopeful about the future because, in our view, a small proportion of television drama already performs some, if not all, of these functions. But we are concerned that the predominance of poor programs and the prevalence of violence may elicit damaging emotional responses in millions of viewers.

In these pages, we have articulated our observations, ideas, and intuitions about television. We examine and appraise a sampling of the many studies of television and its effects, and we draw together recommendations and questions which we think deserve broader attention and further study.

Many of our psychological formulations are speculative; definitive evidence for or against them is not available. Most previous research has focused on short-term viewing effects and is therefore of limited usefulness in assessing long-term effects. But the influences of cumulative television viewing throughout the life cycle are in any case so intermixed with other factors affecting individual development that thoroughly disentangling the specific effects of television viewing is virtually impossible. To have limited our discussion to statistically substantiated objective evidence would therefore have meant omitting pertinent reflections on a number of the more salient features of the television phenomenon.

During the course of our work, the role of violence and commercials directed toward children on television received a great deal of public attention—from the Surgeon General, the National Parent Teacher Association,[3] Action for Children's Television, the American Medical Association, Congress, the public at large, syndicated columnists, and the broadcasters themselves. As this work was going to press, the National Institutes of Mental Health issued a report which summarized ten years of further scientific study on the relationship of television viewing to behavior, updating The Surgeon General's Scientific Advisory Committee Report of 1972. In particular, the NIMH review noted that intervening research had established evidence for a causal relationship between excessive violence viewing and aggression well beyond the level of speculation.[4]

Although the pressure of these collective opinions may alter the patterns of current television programming—by the reduction in video violence, for example—our central concerns remain applicable. All of us, whether parents, teachers, legislators, psychiatrists or television creators, must attend continually to the psychosocial process and outcome of the interaction between the child and the program. We all need to have identified the significant effects of viewing, especially over time. We all need to know how potential harm can be reduced or eliminated. And most of all, we need an augmentation of the personal and social benefits of television.

We consider our own work to be a special mission within a broader quest for a continuous, informed appraisal of the many influences affecting our future. Thus, we address this report to the viewing public, to parents, to the commercial sponsors of television programs, to television writers and producers, and to all people who, like ourselves, are struggling to help ourselves and others deal with the complexities of this changing world.

REFERENCES

1. Wells, H.G. THE OUTLINE OF HISTORY. Garden City, N.Y.: Garden City Books, 1949, Vol. 2, p. 1198.
2. Eisenhower, et al. TO ESTABLISH JUSTICE, TO INSURE DOMESTIC TRANQUILITY, Final Report, National Commission on Causes and Prevention of Violence, Wash.: United States Government Printing Office, 1969, p. 206.
3. National PTA Television Commission, THE EFFECTS OF TELEVISION ON CHILDREN AND YOUTH. Chicago: National Congress of Parents and Teachers, 1977.
4. National Institute of Mental Health, TELEVISION AND BEHAVIOR: TEN YEARS OF SCIENTIFIC PROGRESS AND IMPLICATIONS FOR THE EIGHTIES, Vol. 1: Summary Report. Washington, D.C.: US Department of Health and Human Services, 1982.

1

THE MEDIUM: TELEVISION IN AMERICA

CAPACITY AND POWER OF THE MEDIUM

Television is a phenomenon unique in human experience. Difficult to comprehend fully because of its enormous scope, television appears to have broad implications and profound effects on millions of human beings. Television provokes a wide range of reactions, seldom neutral and mostly strong. Reacting to television, people feel distress, enthusiasm, hope, anger, and pleasure, often in rapid sequence. Is television boon or bane to mankind? Or both?

Television is, of course, many things: a source of entertainment, information and knowledge, an opportunity for vicarious experience. It is a fascinating combination of diffuse influence and focused messages. In our electronic era, it is the major and sometimes the sole source of information for millions of people. Television teaches— for better or worse—a great deal about human relationships, roles, social behavior, and society itself. The ubiquity, availability, and diversity of scheduled programming, as well as its huge audiences all over the world, combine to make television a major social force that requires assessment. How much this gigantic onslaught is absorbed, by whom, and with precisely what effects are problems whose pertinence and complexity are matched by their obscurity. Especially important to consider, in our judgment, is the likelihood that television's effects on the viewer, especially a younger one, intensify through prolonged exposure over time, especially if television viewing weighs heavily in the balance of experience.

The ultimate effects of television are elusive because the vast audience of television viewers is so varied and because the steady trends of cultural homogenization tend to mute consequential differences. Moreover, some effects of television on people are indirect; television's powerful influence on the economy has obviously affected viewer and nonviewer alike. Data about the effects of television, of particular programs, or of characteristics of the medium, range from citations of numbers to arguments about individual cases whose dramatic nature implies but does not verify conclusions. Examples of its positive effects are apparently outnumbered by examples of its tendency to produce harmful effects, such as passivity in chronic viewers, violence in the violence prone, and mediocrity in the masses. On balance, our own view tends to be negative, like that of many other professional people and members of the public. We do, however, recognize that television also can be and has been a source of substantial benefit to viewers of all ages.

At an immediate level, the great value of television as a source of knowledge has been repeatedly demonstrated. From specially designed programs, children have learned numbers, time telling, the alphabet, an expanded working vocabulary, cause-and-effect relationships, and a great variety of other information. In the public schools, television has brought the students quantities of new information from new sources in new and engaging formats for almost every subject in the curriculum.

In another positive role, the magic carpet of television has brought the outside world to a great many individuals confined to the house, to the bed, to an institution, or to a jail cell. Through television, such individuals have vicariously experienced places and events in the world which would otherwise be forever unknown. Television has also played this role for people who are not confined.

Singer has noted that television has played a role in opening new intellectual horizons, adding to the general cultural enrichment of the society.[1] The marvel of television, coupled with earth-orbiting satellites, now makes the entire human community into a potentially simultaneous audience, a fact with many positive implications. The mutual experience of shared events can provide a basis for collective identification, empathy, and cooperation. Man-made problems, such as sea, air, and soil pollution, energy needs, and waste disposal, and natural disasters, such as disease, famine, drought, and earthquake, are rendered more comprehensible and thereby less remote as television depicts their worldwide impact. With a major change in programming emphasis, even the devastation of a nuclear war might be made less likely. Televised celebrations of human activity in the Olympics, of human creativity in art and music festivals, of human achievements in scientific and academic fields, and of compassion in human personal relationships can diminish ethnic, tribal, and ideological divisiveness.

THE UNIQUENESS OF TELEVISION

From those who seek to counter television's critics, one often hears the argument, "Films have used 'excessive' violence for years. Fairy tales are riddled with it. Stereotyping figures appear in all kinds of dramatic presentations—films, radio, the theater. Even comic books were thought to be bad for children. What's so different about TV?"

What's so different? The answer is plain. Television is ubiquitous and continuously accessible and, in this regard, unique. Television is viewed in almost every home, by members of every social and ethnic class and every age group, by the preliterate and the senile, by the prisoner and the school child, by the executive and the laborer. Because it is ubiquitous, television's effects, both positive and negative, are far more significant than if it affected only small numbers of people in isolated sectors. Coupled with the intensity of viewing by a substantial part of the audience and the extended periods of time over which these audiences view television, the size of the audience deepens our concern for their psychosocial health and development.

By 1980 nearly 98% of the homes in America contained at least one television set.[2] In 1979, the average television household had a set on for an estimated seven hours and 22 minutes each day.[3] The television set has become the secular altar in the living room, family room, den, bedroom, kitchen, and even the bathroom of the home. For middle-class American families, television costs little to buy and virtually nothing to run.* Requiring no skill of the operator, it is accessible even to people who are physically handicapped or bedridden. It runs continuously and provides choices, albeit limited ones.

Television has the largest global audience ever assembled simultaneously and regularly for any experience, including church. Well over half the people on earth are said to be directly exposed to it. Television sets now outnumber telephones (364 million compared to 360 million).[4] Precisely because of these dimensions, television is different from any other form of communication. It is, in our opening words, unique in human experience.

THE ECONOMICS OF THE TELEVISION INDUSTRY

In any discussion of television and its impact, one must keep in mind the economic size of the television industry. According to *Broadcasting Magazine,* advertisers spent over eight billion dollars on television advertising in 1979, an amount that has increased steadily each year.[5] The amount represents hundreds of thousands of jobs in the advertising industry, in the television industry, and in the businesses associated with and supportive to them.

The broadcasting and motion picture industry is one of the most profitable in the world. In 1978, this industry showed the highest rate of return on sales (9.6%), on stockholders' equity (21.6%),

---

* In many parts of the world and in poorer parts of the United States, the cost of a television set is staggering in proportion to the resources of families. That they will nonetheless elect to buy a set illustrates the astonishing importance of owning a television.

"...the secular altar..."

Photo, courtesy of Bill Owens

and on total return to investors (33.34%) of the Fortune 500 companies.**[6] Although it would seem reasonable that such a profitable industry could afford to make choices on other grounds, decisions made in the television industry are based on the same premise as they are in other industries. Choices are based on what will produce the maximum profit. The larger the audience, the greater the demand for advertising time and the higher the fee that advertisers will pay for a minute of television time. This may mean spending anywhere from $110,000 to $300,000 per minute for a prime time commercial, with an average of $160,000.[7] By comparison, *Variety* estimated that the cost of producing an entire thirty-minute, prime time show may range from $210,000 to $265,000 and a sixty-minute show from $400,000 to $450,000.[8]

To produce the type of program that will garner the maximum profits, the industry follows the formula described by Bob Shanks, a television network vice president. "Program makers are supposed to devise and produce shows that will attract mass audiences without offending those audiences or too deeply moving them emotionally. Such ruffling, it is thought, will interfere with their ability to receive, recall, and respond to the commercial messages."[9] The bottom line is what will sell or, put another way, what the masses are willing to buy. How well television critics or educators receive a program makes little difference. Unless the program can attract large audiences and score in the ratings, it will soon be off the air.

Based on the billions spent each year by television advertisers, we can assume that advertisements do work. Each year, advertisers continue to spend more and more money, which in turn brings the networks more and more profits. Notably, of the profits, 25% come from the 7% of programming directed toward children.[10] Moreover, as Peggy Charren of Action for Children's Television has stated, "children are the objects of a $400-million-a-year advertising assault... It is obvious that advertisers have not made this kind of investment to attract... (a few pennies) from a child's allowance.

** All-industry averages were: 4.8% return on sales, 14.3% return on stockholders' equity, and 7.16% on total return to investors.

In fact, the parents are the purchasers, and the children assume the role of surrogate salesmen, the advertisers' personal representatives in the home."[11]

The fact that the television programming and production process is so deeply embedded in a commercial context places sharp limits on the kinds and degrees of changes which might be introduced. Writing in the *New York Times*, William V. Shannon stated, "TV can never be substantially reformed as long as it remains solely in the hands of businessmen preoccupied with packaging the largest possible audiences for sale to hardsell advertisers."[12]

Because television is a mass-audience medium, judgments about its content are commonly based on ratings. Though many have deplored the ratings' influence, the industry retains them as central in decisions about what is shown and when. Although present trends are not encouraging, advances in television technology and new developments in the means for distributing television programming will substantially increase the number of choices of what to watch. Expanded cable facilities, videodiscs, satellite receiving centers, and low-power neighborhood TV stations may in time release viewers from the stranglehold of the programming monopolies which dominate commercial television and from the current rating system. Two-way cable systems that allow viewers to register opinions about various agenda items may ultimately enable the viewers to have great influence over program offerings.

How responsive any industry should be to the needs of society is an important question of public policy, larger than our special professional perspective can embrace. Television, in our view, certainly has as much or more responsibility as the automobile, energy, or building materials industries. We believe that at minimum those involved with the production and marketing of commercial television programming should become more aware of its effects, both positive and negative, and that they should share the obligations of other professions to make continuing efforts to improve the quality of the impact of their products.

TELEVISION AS A PSYCHOSOCIAL ISSUE

Television is now a major source of ideas, values, and tastes in our society, whether by design or by chance. The information (or misinformation) it conveys affects how we organize our lives, how we raise our children, when and what we eat, what we do or do not buy, what we think of law enforcement, policemen, doctors, and others, what issues we believe are important (or even know about), what problems we regard as important to solve and how we ought to go about solving them, and countless other aspects of living. Ostensibly products of our culture, television programs also shape the culture. They have a substantial impact on the formation of viewers' personalities, values, and aesthetics, and their ideas about cause and effect, reward and punishment, power and authority, and the human prospect.

Because the bulk of television programming superficially imitates reality, it appears to be an adequate and accurate mirror of reality. The information it conveys is, thus, commonly taken as reliable, and the messages are believed to be legitimate. What we see on television can easily become the standard by which we compare and judge what is actually happening in our lives (often unfavorably). Two observers, Mankiewicz and Swerdlow, have suggested that the intense "realism" of television presentations has caused an inversion of consciousness and that nothing is "really real" unless we have seen it first on television.[13] It is as if reality has become less authentic than the television image.

Television gives us a far greater range of vicarious experience than we could ever expect to have in reality. Indeed, Mander has written, "America [has] become the first culture to have substituted secondary, mediated versions of experience for direct experience of the world. Interpretations and representations of the world [are] being accepted as experience, and the difference between the two [remains] obscure to most of us."[14] The result, he suggested, has led to a "new muddiness of mind" having serious consequences.

Via its billions of images of reality which are assembly line produced, distorted, and formula ridden, projected for many hours, months, and years on end, television may affect our view of what is real more than any other social influence. Television scenarios, for example, radically simplifying human personalities and human relationships and showing us pruned patterns of attitudes, behavior, and ways of speaking, tend to encourage viewers to expect real-life solutions that are as neat, complete, and quick as those on the screen. That even occasional exposure may have powerful effects on some individuals seems clear. But how much stronger and how much more consequential must be the influence of prolonged viewing throughout early childhood, school age, adolescence, and adulthood?

Assumption of greater cumulative effects is in accord with our understanding of personality development, acculturation, and the motivations of behavior as derived from our clinical observation and experience. One such example from a psychiatrist's practice illustrates how television drama can have negative effects on children through their parents:

> A young wife traced her husband's increasingly frequent battering of their two small children to a period of greatly increased television viewing by both parents and children. Prior to this period, they had watched selected programs on a black-and-white set for just a few hours a month. With the purchase of a color set, their viewing increased over a period of three years to about 30 hours a week. And the programs they now watched indiscriminately included much violence. Concurrently, the emotional climate of the home changed from "happy-go-lucky" to "unhappy," with adverse effects on all the members of the family. In particular, the wife noted a worsening of her husband's "short temper." Previously it had never led to more than screaming at a crying child or banging his fists on the crib. More recently, however, the father had inflicted serious injuries on his children.
>
> In counseling, the parents learned how the husband's frequent

severe beatings by his own father during childhood had set him up to punish his children in the same way. When they were asked by the counselor to consider why this previously latent tendency was now producing actual injury to his children, the wife suggested that the change in their television-viewing habits might have contributed to the change in the emotional atmosphere of their home.

With reluctance, the husband agreed to a reduction of their viewing (including his own) to two hours a day, with exclusion of violence except for that which may be integral to quality programs. The wife reported that, although the home atmosphere still was stressful in some ways, after six months on this reduced television diet some joy had come back into their lives. "We laugh more. We're kinder to each other. The children don't fight continuously. And Al hasn't hurt them since we made the viewing change."

The psychiatrist's opinion was that the constant modeling of impulsive, violent discharge of tension on television mobilized the man's identification with his own battering father and weakened his already shaky defenses against his own violent impulses. Whether the reduced viewing, the therapeutic counseling, or some other influence brought about the beneficial changes, we are reminded by this vignette that television is a prime example of an experience that is intense, especially arousing, and frequently repeated. As a consequence, it may well have effects on other aspects of a person's life.

Another example illustrates the complexity of direct effects on children when they view emotionally stirring television drama:

> A four-year-old girl, the daughter of broadly tolerant white parents, awoke from a nightmare screaming, "Mommy! The black people—help me!" The next day she told her mother, "Black people are bad." Pressed for further explanation, she said, "I saw it on *Roots*," and added, "The white people are good. They wouldn't whip anyone who wasn't bad." Perplexed, the mother wondered how her child could have gotten an impression about "good" and "bad" people so much at variance from what the producers had intended.

> By analogy, when the child saw a (white) person whip a (black) person on TV, she concluded from her own experience of being spanked that the (black) victim must have been bad—and that such a bad person might also punish her.
>
> Acknowledging that she spanked her child—"...but only when she's bad," the mother agreed that perhaps the idea of being spanked conveyed just that message: it is what "good" people do to those who are "bad."

The viewing of an episode is not the only relevant factor, nor is the influence of a particular episode universal. Several older siblings of the child in the example saw the same episode of *Roots*. Although they, too, were spanked for bad behavior, they did not have the same reactions.

The example also illustrates that the process and effect of television viewing are a complex interaction between the psychological state, particularly the emotional and cognitive maturity, of the child, the content of the program, the viewer's environment, and the immediate situation at the time of viewing.

Comstock divides the complex of interacting factors into three broad categories:

1. person-related factors: age, sex, personality and level of cognitive development;
2. stimulus-related factors: program content and the way it is presented; and
3. environment-related factors: the characteristics of both the specific viewing situation and the person's living conditions.[15]

Our report focuses primarily on the first two of Comstock's three categories. We examine the person, the developing child, and discuss some common assumptions about how television interacts with children's developmental needs. With the child's needs as the backdrop, we then consider the major properties of several of the types of dramatic programs that children watch. As we present our recommendations, we touch briefly on environment-related factors.

REFERENCES

1. Singer, Jerome L. (Ed.) THE CONTROL OF AGGRESSION AND VIOLENCE, NY: Academic Press, 1979, p. 19-60.
2. Adler, R.P., Lesser, G.S., Meringoff, L.K., Robertson, T.S., Rossiter, J.R. and Ward, S. THE EFFECTS OF TELEVISION ADVERTISING ON CHILDREN. Lexington, Mass.: Lexington Books, 1980, p. 26.
3. Shannon, Martin J. "Business Briefs" *Wall Street Journal,* 3 April 1980, p. 1.
4. International Institute of Communication. *Intermedia.* London: International Institute of Communication, 1977.
5. "TV in '78: The Biggest Net Yet." *Broadcasting,* 30 July 1979, p. 38 ff.
6. Dworkin, Peter. *Fortune,* 7 May 1979, p. 290-291.
7. Coates, Colby. "Prices for 1979-80 Prime Time Network TV Programs (per 30-second unit)." *Advertising Age,* 10 September 1979, p. 98.
8. "1979-1980 Network Primetime Season at a Glance." *Variety,* 12 September 1979, p. 72.
9. Shanks, Bob. THE COOL FIRE: HOW TO MAKE IT IN TELEVISION. New York: W. Norton & Co., 1976, p. 98.
10. Rothenberg, Michael. "Effects of Television Violence on Children and Youth." *Journal of the American Medical Association,* 1975, *234* (10), 1043-1046, Dec. 8.
11. Charren, Peggy. "Should We Ban TV Advertising to Children? Yes." *National Forum,* 1979, *69* (4), 15.
12. Shannon, William V. "The Network Circus." *New York Times,* 3 September 1975, p. 37.
13. Mankiewicz, F., and Swerdlow, J. REMOTE CONTROL—TELEVISION AND THE MANIPULATION OF AMERICAN LIFE. New York: Times Books, 1978.
14. Mander, Jerry. FOUR ARGUMENTS FOR THE ELIMINATION OF TELEVISION. New York: Morrow Quill Paperbacks, 1978, p. 24.
15. Comstock G., Chaffee, S., Katzman, N., McCombs, M., and Roberts, D. TELEVISION AND HUMAN BEHAVIOR. New York: Columbia University Press, 1978, p. 264.

2

THE AUDIENCE: CHILDREN AS VIEWERS

It is no secret that today's children watch a great deal of television. In homes with children, the television set is on an average of 59 hours 50 minutes a week, as compared to 42 hours 57 minutes weekly in households without them.[1] How much television children watch varies with age and race.[2] One study noted, "Data... reveal that preschoolers are watching 27 hours 4 minutes per week on the average."[3] Lyle and Hoffman found that nearly all the first graders in their California sample watched television every day, and most of them watched for a least two to three hours; well over one-third watched four hours or more.[4] Although the average time watched reaches 3.7 hours per day by age 12, or about 27 hours per week, and slowly declines thereafter,[5] about 25% of the sixth and 10th graders had watched at least five-and-a-half hours of television on a given school day. This viewing extended well beyond the family viewing hours as usually defined, and the patterns of programs selected did not match the programming category of "children's television." By the first grade, 40% of the children's viewing was devoted to programming classified as "adult."

It has been variously estimated that by the age of 18, a typical teenager has spent considerably more time in front of a television set (15,000-20,000 hours) than in school (11,000 hours). In the course of that watching, he or she has seen some 13,000 killings, more than 101,000 violent episodes, and between 350,000 and 640,000 commercials.[6] That amounts to some 20,000 commercials each year, or more than three hours of advertising each week.[7]

Although massive exposure *per se* does not automatically make television a powerful teacher, children do learn ideas, values, behaviors, roles, attitudes, emotions, and responses from television.[8] Children acquire some information from programs inten-

17

tionally designed to teach, such as those produced for the classroom and such broadcast offerings as *Sesame Street* and *Mister Rogers' Neighborhood*. But children also acquire information from programs which are primarily intended to entertain.

Many studies confirm that television affects young persons' attitudes and expectations about the nature of the world and the people around them and that it can also significantly change children's attitudes. Experiments have clearly demonstrated that young children replicate the behavior they observe on television, including aggressive acts and such socially beneficial behavior as cooperation, rule obedience, and verbalization of feelings. What remains to be done in the realm of research is to specify the factors that influence such learning and the conditions and contingencies that affect whether or not and how a person will transform and manifest the facts, feelings, ideas, and attitudes he or she acquires from television.[9]

NEEDS OF THE DEVELOPING CHILD

Critics of television, concerned about the deleterious influence of television on young viewers, are inclined to speak of children as if they were *tabulae rasae*, innocent creatures molded by and entirely at the mercy of nefarious television shows. Such a view ignores the basic psychological realities of children. Children possess feelings and drives that are relatively uninhibited. Without any television at all, they have their own strong feelings of anxiety, anger, rivalry, jealousy, possessiveness, self-centeredness, fear, shame, and guilt. Adults who exaggerate the image of the helpless child corrupted by external influences, such as the media, obscure the complex interaction of individual and environment.

In a study of British children, Greenberg found they had a number of their own reasons for watching television. Some, especially teenagers, wanted to acquire new information. Others sought distraction and wanted to avoid boredom. Children also viewed, they said, for relaxation and, paradoxically, for excitement. Some found companionship through viewing. And some watched simply to feed their "television habit."[10]

Halloran observed that what children do with television may be as significant as what television does to them.[11] He observed that those children least able to cope effectively with their social environments—those with lower intelligence and less stable family and peer relationships—more frequently seek out escapist programming on television. He suggested that while such programs may help to ease the anxieties threatening the child, they also appear to deter solution of the very problems that initially motivated the child to search for relief. For example, the difficulties of frustrated and maladjusted children are intensified when violent content reinforces existing behavior.

Schramm, Lyle, and Parker correlated their findings on children's viewing patterns with a number of individual characteristics. Notable among these was intelligence: at early ages, the brighter children viewed more, but later they turned away from television. Intellectually slower children, on the other hand, became heavier viewers. The amount of viewing was also correlated positively with family troubles and unsatisfactory social relationships.[12] This is of particular interest in light of the many changes families are undergoing.

The changing family environment

The changing nature of the nuclear family has influenced the role that television viewing plays in the life of the family and in the growth and development of the children. Less and less often are extended family members available to help the young mother. Extraordinary geographic and social mobility has weakened both family ties and social support networks.

In sharp contrast to the agricultural family, urban fathers are intermittent inhabitants of the household and often tend to see the home as a place for rest and recreation. He may view his wife as a support to his career enhancement. The family they share may be a normative feature of modern middle-class living, without necessarily being a cooperative venture in child rearing.

In recent years mothers with school-age children have entered the job market in steadily increasing numbers. In two-paycheck families, family interaction may diminish, and the children frequently turn to the television set for companionship and stimulation.

Today's numerous single parents confront substantially heavier child-rearing burdens than couples. A single parent must combine maternal and paternal parenting functions into one, with the parenting role still further complicated by the need to earn a living. When the single status of the child-rearing parent is a consequence of divorce, the psychological support for both parent and children may diminish sharply, along with physical support. The resulting isolation and reduction in resources make an intimate and stimulating family environment considerably more difficult to achieve.

In such a context, television may play a prominent role. It fills up empty time and space, offers stimulation and amusement, and serves as a dependable baby-sitter. And further, television is an ever-responsive but undemanding companion. Thus, television is inevitably used to meet psychological needs for which it is ill suited. It substitutes for the intimacy and engagement which are normally part of the parent-child relationship. Especially for the many single parents whose own parents and siblings are inaccessible as surrogates or baby-sitters, television substitutes for the companionship of the absent parent(s). And, with the adults absent, working, or busily preoccupied in the house, the children are left alone and unobserved to watch whatever the television has to offer.

The psychological process of identification

There are many psychological processes at work as the child moves from infancy to adulthood. One process which is of particular importance in relation to children and television is that of identification. Identification is the process whereby a child internalizes certain aspects or qualities of a person and makes them part of his or her own developing ego. Identification usually occurs in the context of a positive relationship and is growth promoting. It may, however, be defensive, as for example when a child identifies with the aggressive qualities of a feared individual to protect himself or herself from the anxiety and helplessness experienced with that individual. In this instance, identification may reduce anxiety, but at the price of potentially impaired future relationships and less adaptive ego skills. The identification process is complicated and deeply affected by many important factors. Parental models and parent-child relationships, the child's age and psychological level of development, and the presence and adequacy of environmental supports all influence the process.

The interactions of the many influences of identification are so complex that careful study of the importance of any one influence is very difficult. Our inferences about the role of television drama in the identification process are limited by our meager knowledge, underscoring the urgency of designing and funding further pertinent studies. Still, although we do not at this point know exactly how influential television models are on the identification process in the growing and developing viewer, the antisocial, narcissistic aspects of some dramatized television characterizations differ so much from models of normative, healthy personalities as to justify considerable concern.

The identification processes which affect the direction of a child's growth and development are influential both at the younger ages and into adolescence and beyond. At each age level, examples of desired and admired behavior are consciously recognized and unconsciously incorporated as models to imitate. Young children have heroes and heroines and often act their parts in make-believe play. Teenagers consciously select figures they admire to copy. Young adults are often profoundly affected by significant nonfamily members (youth leaders, teachers, employers, older friends, and so forth). Unconsciously, they adopt such figures as models for their evolving personality, even through adulthood. Individual personality and temperamental characteristics, such as self-concept, impulse control, developmental level, goals, and aspirations, affect how influential such figures are and why they are chosen. Examples of strength have special meaning for the youth who feels weak and without the resources to gain strength. Examples of competence, power, skill, or cleverness lend themselves as models to the individual who feels great inadequacy. The process is therefore selective, since it reflects the internal needs of the individual. By its nature, the process is self-focused and gives little weight or attention to the needs or interests of others.

When models of behavior or character suggest the acceptability of an infantile, narcissistic posture, they exemplify means of attaining self-esteem which are essentially antisocial, or appealing to the selfish interests of the individual without concern for the rights or feelings of others. Such models are antagonistic to the healthy psychological growth of the individual and to the society in which such concerns are crucial.

The likelihood that a child will identify with such models depends on his or her need or readiness to seek such models of behavior, as well as the intensity and verisimilitude of the televised experience. The quantity of the exposure may also be quite important, as, of course, are the child's age and developmental level. Occasional presentation of heroes whose behavior is selfish, impulsive, or antisocial will have quite a different impact on an individual viewer than will such presentations endlessly repeated.

In this consideration of identification, we must concern ourselves not only with the presence of injurious models but also with the absence of beneficial ones. The importance of admirable models to the adolescent is well known, as is the adolescent's quest for enduring patterns of justice and kindness. Adolescents seek exemplars who demonstrate confidence, competence, and success. When unjust or cruel examples are offered by otherwise attractive models, growth towards socially desirable patterns is impeded or blocked, and emulation of the antisocial example can occur. On the other hand, when programs offer positive ethical concepts, expressed by attractive characters whose behavior satisfies the hunger for a believable image of the just and kind person, such growth in adults as well as adolescents is facilitated.

An example of the latter is *Star Trek*. This series has attracted an intensely devoted worldwide following of adolescents and young adults. They speak with admiration about protagonists who prefer peaceful solutions and wherever possible rely on reason to resolve emotional struggles. Roles and behavior are sufficiently clear to prevent confusion of fantasy and reality. Even though the story is set in a world of the far distant future, it is not so artificially purified as to be misleading or alien. In this world, conflicts are resolved by intelligence, mastery of challenge, good will, humanism, and altruism. These are values which adolescents share and believe in with passionate intensity and on which human survival depends. Unfortunately, there are few programs which exemplify these values in an engaging way.

The unique interplay between the process of a child's identification and the idealized roles and models so typical of drama, and television drama in particular, is among the more important areas for future research. For the moment, given our lack of experimental corroboration, our speculations must stand on their own as the informed opinions of clinicians who study human psychological development.

STAGES OF PSYCHOLOGICAL DEVELOPMENT

As already noted, the age and level of development of the child in question are paramount mediators in the effects of television viewing. Since childhood is a lengthy period characterized by growing capacities and changing developmental needs, it is essential to consider the television viewing experience in light of the child's particular age and stage of development.

Infancy and toddlerhood

Television has its most significant effects when the child's interest in it and exposure to it are at a maximum. For the average infant and toddler, television viewing is probably not a significant experience. Visual, auditory, and language skills have not developed to the point where television has meaningful appeal. Moreover, television certainly cannot draw the infant from the much more intense attraction and attachment to his or her mother (or primary caretaker). Except when an infant has been deprived, the child at this age is predisposed toward interaction with a responsive human being. These interactions of mutual cuing, exploring, expressing, and responding are precursors of the child's future relations with siblings and other family members and of later ones with peers. The child's capacity to learn about and relate to other people and the world depends very much on how gratifying and reinforcing were the early interactions with his or her mother.

Infants also learn by interacting with inanimate objects, receiving gratification and reinforcement from the results of their own physical actions. A four-month-old child, for example, shows sustained interest in an overhead mobile to which his or her wrist is attached with yarn. As the body moves, the mobile moves. The experience of having produced the motion is an early example of learning through interaction and reinforcement.

When an infant or toddler is deprived of sufficient interactive learning opportunities, particularly with a caring adult, the young child may seek substitute sources of gratification and stimulation. For the infant up to the age of 18 months, this may initially result in increased sucking and crying. For the toddler (18 to 36 months), it may result in body rocking, excessive eating of sweets, abnormal immersion in a relationship with a pet, or too much watching of television.

At the toddler stage, television may assume the role of a transitional object, akin to the familiar teddy bear or favorite blanket. The auditory and visual stimulation the television gives makes the toddler feel comforted and less alone. Whether this exposure is positive or negative probably depends more on the quality and availability of caring adult figures than on the nature of the television programming itself.

An individual's style of learning and many basic personality traits are established in the earliest years. Those children who have had to use television as a substitute for learning and human interaction are more prone to develop passive learning styles.

Preschool years

As the child moves into the nursery school years, visual, auditory, and language skills have developed to the point where he or she will find meaning in television programming. Cartoons, because of their simplicity, have particular appeal to preschoolers.

Of particular importance as developmental issues for the preschool child are the development of autonomy and self-control and the mastery of strong feelings, particularly anger. These feelings can threaten the developing self-confidence of the child, as well as relationships with family members.

Between the ages of three and five, the normal child is increasingly active, inquisitive and eager to learn about everything. The predominant television viewing pattern for children of this age is a readiness to watch—to watch other children, and to watch what they watch, even though much of what they see will not make much sense to them. If the child's opportunities to learn are severely restricted, the outward search for new experience may be blunted or even stopped altogether. Television may be the only available source of novelty and interesting stimulation. Ultimately, television viewing can become an important substitute for active engagement with the environment, making it hard at later ages to use the interactive approaches to learning required in school and expected in social relationships. The undue reliance upon television at an early age may establish or simply reinforce a habit of passive dependence on the environment for all manner of perceptual and intellectual stimulation and amusement. When this occurs, the preschooler may fail to accomplish key developmental tasks in the areas of assertiveness and independence. These developmental failures set a child up for later trouble in the school-age and adolescent years.

Greenberg and Dervin found that the role played by television in children's lives reflected significant differences in socioeconomic status and race.[13] Within their sample, one which did not represent the entire population, they noted a strong inverse relationship between the amount of viewing and the family income for whites. The same was true for blacks, but the relationship was far less strong. Amount of viewing was greater for black children than for white, independent of family income. Black children more often acknowledged that a motivation for watching was to learn things, and they more frequently believed that television accurately portrayed real life. Children of lower-income families, both black and white, devoted more time to television, received less guidance about television viewing from parents, and expressed greater belief that television accurately depicts real life than did their higher-income peers.

It is clear that inadequate housing conditions, street violence, the lack of alternative child care facilities, and other factors associated with poverty and ghetto living foster this use of television. The effects of economic disadvantage and minority status need to be taken into account in evaluating viewing patterns.

School age

Most of the research and writing about the impact of television viewing has focused on school-age children. During the school years, the child makes significant cognitive gains, affecting the capacity to understand and to learn from television programming. Particularly important is the growing capacity to make sense of what he or she sees or hears. Cognitively, he or she moves during the school years from fragmentary notions to progressively integrated concepts, from isolated events to concepts of sequence and causality, and from interest in concrete objects to interest in abstract ideas.

During the school years, the child's capacity to relate to peers steadily increases. As part of this process, the child begins to develop a sense of personal values and morality. An examination of the basis for moral judgments of young children suggests that beginning at about age seven the child shifts from "objective morality," in which the source of values is outside, to "subjective morality," in which the child begins to define his or her own values. These findings are consistent with classical developmental theories, particularly those of Piaget.[14] With increased cognitive capacity and the emergence of subjective morality, the child learns to comprehend the idea of motives and to infer logic.[15]

Illustrating the impact of age and cognitive development on the extent and kind of learning, Collins found that children as old as second and third graders often pay attention to and remember fewer or different aspects of programs than expected.[16] The result is often an incomplete or distorted view of the people and the events they see and learn about on television. Comprehension of significant aspects of story material is both more limited and more fragmentary than expected.

School-age children seem to remember a much smaller proportion of the essential information that is explicitly presented in single scenes than do older children, adolescents, or adults. Second graders recalled an average of only 66% of the scenes that adults had judged essential to the plot; fifth graders recalled 84%; and eighth graders recalled 92%.

Their comprehension is also fragmentary. Even when children do recall significant content from essential scenes, they may fail to grasp the interscene relationships. Adults readily infer that what happened in a scene early in a program is pertinent to some later scene, but up until age nine or 10 most children are rather poor at inferring causal connections. Second graders remembered an average of fewer than half (47%) of the items identified by the adult judges as inferentially significant; fifth and eighth graders remembered 67% and 77%, respectively. Younger children apparently do not take advantage of the information provided indirectly by temporal order.

The fact that younger children operate in this fashion makes it likely that they will not comprehend that violence in a given scene may have been *caused* by an event in an earlier scene. While such young children will frequently remember the violent action in a television program, they will only infrequently understand its links with motives and consequences. Their lack of appreciation of adult motivations foreign to their experience adds to their developmental limitations and makes school-age children particularly susceptible to negative aspects of dramatic television. Futhermore, children who are emotionally aroused by an upsetting scene may be deprived of the tension release provided by further developments in the plot if these developments are too subtle for children to follow. Their exposure needs to be carefully controlled.

With the entry into school, children begin the long process of moving further away from home and family. The importance of school, the peer group, after-school activities, hobbies, and the like steadily increases each year as the time spent at home and with the family slowly decreases.

Although family relationships remain critically important through these years, the quality of these relationships changes. There is increasing respect and allowance for individual desires as dependency needs become less and less prominent. For the school-age child, therefore, television viewing can serve as a convenient alternative or excuse which distracts the child from the necessary, occasionally painful, but ultimately rewarding process of separation.

Adolescence

As a child reaches adolescence, with its accompanying cognitive development, he or she can increasingly understand and interrelate multiple causal factors and appreciate the many qualities and characteristics of a single object. The ability to distinguish fantasy from reality is much more refined. With an increasingly autonomous sense of self and personal values, the adolescent less readily accepts things at face value. Other sources of information and experience with the real world increase dramatically. As a result, some of the concerns expressed earlier about television's effects become less intense.

The developmental tasks of adolescence draw a person further away from the family and into even more peer-related activities, allowing less and less time for television viewing. During the adolescent years, separating from family, increasing one's sense of identity and independence, relating more with peers, and developing the capacity for intimacy are the areas for psychological growth. In the course of adolescence, given an adequate range of interests and pleasure in academic and social success, the adolescent finds television a less significant experience.

Some adolescents, however, encounter emotional difficulties handling the normal stress of this developmental period. For them television can be a refuge. By staying at home and watching television, frequently in isolation, the adolescent can avoid dealing with intimacy with others, peer relations, and the separation from the family. A heavy "television habit" in an adolescent, therefore, is of much greater significance and concern than it is in a school-age child.

The pivotal issue for adolescents is the beginning of the development of identity, a task which will be a lifelong process. The identification process, set in motion initially through earlier developments, reaches a critical point in adolescence. In relation to television viewing, the need of the adolescent for role models and the proclivity to imitate them have both positive and negative possibilities.

On the negative side, the role of television as a possible trigger to violent impulsive behavior in some individuals has gained much attention over the past several years. We are all too familiar with the most highly publicized episodes in which a television program about a broomstick rape, a tower sniper, a bank robbery, or the burning of a "hobo" with gasoline is followed by real-life, imitative acts. The individuals involved in these acts, it must be emphasized, had severe, preexisting problems with behavior and impulse control; thus, the interaction between them and the television program is quite complicated. A certain minority of adolescents do have severe difficulties in impulse control, tendencies for direct discharge of impulses, and faulty or incomplete development of personal values and morality. Coupled with their normal adolescent physical and intellectual capabilities, these personality traits may make them more vulnerable to the violent, impulsive behavior seen on television, and result in direct imitation of televised actions.

The case of the adolescent differs from that of the preschool or school-age child. Children can and do act out impulses, but this usually occurs safely in the context of play. This does not mean that imitative behavior is not important for the younger children, but that no one usually gets hurt in a make-believe setting. The potentially serious consequence, however, is that in certain vulnerable youngsters this imitative play could set down patterns of behavior to be acted out in real life as the youngsters mature. In contrast, some adolescents (and adults), due to normal developmental tendencies and certain psychological vulnerabilities, are susceptible to television stimuli and thus more likely to respond by acting in disturbing ways.

Television dramas could, however, provide numerous positive models to both the vulnerable and the normal adolescent. Aside from *Star Trek*, however, television offers few programs which portray socially beneficial resolutions of the emotional problems faced by adolescents and which are also relatively free of violence. An exception to this statement is *Welcome Back Kotter*. A group of adolescent boys and girls were observed watching and discussing

Watching "Come Back Kotter?"

Photo, courtesy of Shearer Visuals

the program in weekly gatherings. Members of the group had one common characteristic: they were all nonreaders. They did not even read the sports pages, although all were athletic. Some were probably dyslexic, and others clearly had emotional blocks about reading. For example, one 16-year-old was in conflict over his allegiance to a nonintellectual, successful businessman father, and an intellectual stepfather and a mother who enjoyed reading. Another boy had ulcerative colitis, and sports were his way of proving his vitality.

Each of these young people found a common hero in the charismatic teacher, Kotter, who understands their adolescent conflicts from both a psychological and a social point of view. He conveys a sense of acceptance and legitimacy which adolescent viewers find quite supportive. Viewers describe him as an adult who "doesn't just act like a parent!"

By demonstrating a warm and positive relationship, Kotter provides the conditions which encourage his teenage students toward better and more accepting self-concepts. By identification, the teenagers who watch receive this encouragement, too. As the program illustrates a variety of social and emotional problems typical of the teenage viewing audience, Kotter's positive approach encourages viewers to work their problems out and teaches them useful problem-solving skills.

Summary

As children change, so do their relationships with television. As the foregoing discussion illustrates, the effects of television viewing vary with the child's age and developmental level, with the content of the program, and with the child's particular patterns and circumstances of viewing. As well as being concerned about the differential effects at various developmental stages and under varying viewing circumstances, including program content, we are also concerned about the cumulative effects of viewing television throughout the childhood years. A central psychological process of importance in understanding the effects of television on children is identification, the process of internalizing certain qualities of a person and making them part of one's own developing ego. Television has enormous potential to influence this process.

In the course of public and professional debate about the effects of viewing, some common questions and mythologies have arisen. How television viewing affects play and fantasy, whether it promotes passivity and dependency, and how it relates to loneliness are common themes in the dialogue and debate about children and television. Our following discussion of these issues illustrates the need to be sensitive to both the long-range and the short-range circumstances and effects.

REFERENCES

1. 1982 NIELSEN REPORT ON TELEVISION, p. 7.
2. Liebert, R.M., and Schwartzberg, N.S. "Effects of Mass Media." *Annual Review of Psychology,* 1977, *28*, 141-173.
3. Nielsen, op. cit. p. 8-9.
4. Lyle, J., and Hoffman, H.R. "Children's Use of Television and Other Media." In E.A. Rubinstein, G.A. Comstock, and J.P. Murray (Eds.), TELEVISION AND SOCIAL BEHAVIOR, Vol. 4. Washington, D.C.: Government Printing Office, 1972, 257-273.
5. Adler, R.P., et al. THE EFFECTS OF TELEVISION ADVERTISING ON CHILDREN. Lexington, Mass.: Lexington Books, 1980, pp. 13-28.
6. Steinfeld, Jesse L. "TV Violence *Is* Harmful." *Reader's Digest,* April 1973, p. 37-45; Somers, Anne R. "Violence, Television and the Health of American Youth," *New England Journal of Medicine,* 1976, *294* (*15*), 811-817; Rothenberg, Michael. "Effects of Television Violence on Children and Youth." *Journal of the American Medical Association,* 1975, *234* (10). 1043-1046, Dec. 8.
7. Adler, R.P., et al. THE EFFECTS OF TELEVISION ADVERTISING ON CHILDREN. Lexington, Mass.: Lexington Books, 1980, p. 18.
8. Leifer, A.D., Gordon, N.J., and Graves, S.B. "Children's Television: More Than Mere Entertainment." *Harvard Educational Review,*

1974, *44* (2), 213-245; Comstock, G., Chaffee, S., Katzman, N., McCombs, M., and Roberts, D. TELEVISION AND HUMAN BEHAVIOR. New York: Columbia University Press, 1978, 261-287; Stein, A., and Friedrich, L. "Impact of Television on Children and Youth." In E. Hetherington (Ed.), *Review of Child Development Research,* Vol. 5. Chicago: University of Chicago Press, 1975, pp. 183-256.
9. Comstock, G., Chaffee, S., Katzman, N., McCombs, M., and Roberts, D. TELEVISION AND HUMAN BEHAVIOR, New York: Columbia University Press, 1978, p. 261 ff.
10. Greenberg, Bradley S. "British Children and Television Violence." *Public Opinion Quarterly,* 1974, *38,* 531-547.
11. Halloran, J.D. THE EFFECTS OF MASS COMMUNICATION WITH SPECIAL REFERENCE TO TELEVISION. Leicester, England: Leicester University Press, 1964. (Television Research Committee Working Paper #1.)
12. Schramm, W., Lyle, J., and Parker, E.B. TELEVISION IN THE LIVES OF OUR CHILDREN. Stanford, CA: Stanford University Press, 1961.
13. Greenberg, B.S., and Dervin, B. USE OF THE MASS MEDIA BY THE URBAN POOR. New York: Praeger, 1970.
14. Chandler, M.J., Greenspan, S., and Barenboim, C. "Judgments of Intentionality of Response to Videotaped and Verbally Presented Moral Dilemmas: The Medium is the Message." *Child Development,* 1973, *44,* 305-320; Hebble, P.W. "The Development of Elementary School Children's Judgment of Intent." *Child Development,* 1971, *42,* 1203-1215.
15. Leifer, A.D., Collins, W.A., Gross, B.M., Taylor, P.H., Andrews, L., and Blackmer, E.R. "Developmental Aspects of Variables Relevant to Observational Learning." *Child Development,* 1971, *42,* 1509-1516.
16. Collins, W.A. "Temporal Integration and Children's Understanding of Social Information on Television." *American Journal of Orthopsychiatry,* 1978, *48* (2), 198-204.

3

THE AUDIENCE: SOME QUESTIONS OF EFFECT

HOW DOES TELEVISION AFFECT PLAY AND FANTASY?

Why fantasize?

As clinicians we are especially interested in the relationship of television to the development of imagination in the young child, the effect of television on fantasy, and the possible role of television in causing problems of separating fantasy and fiction from reality.

Fantasy and imagination are important faculties for the developing child. Fantasy allows the child to explore, or try out, all sorts of feelings. Fantasies absorb a child's feelings of rage or jealousy, feelings too dangerous to express in real life. Fantasies may fill in for something lost or missed, or make up for feelings of deficiency. Through imaginary action, a child can try out dangerous or forbidden acts without guilt, shame, or fear. As Bettleheim has suggested, fantasy is more than a simple escape valve for intolerable tensions, anxieties, and ideas. Fantasy is an essential part of healthy personality.[1]

Imagination and fantasy are particularly enhanced in the growing child's use of play. Symbolic games, pretending, and make-believe can all contribute to the ability to fantasize. A number of investigators have suggested that make-believe play is a form of symbolic manipulation and control over realities and experiences.[2] Fantasy is a source of ideas and a means of projecting the self backward or forward in time or space. A vital ingredient in the problem-solving process, the ability to fantasize underlies the capacity to imagine future actions and consequences.

As an important means for reviewing experience, fantasy and make-believe help a child work through a loss or trauma in the same manner that repetitive play helps a child to master disturbing experiences and feelings. Some investigators have reported that children who regularly engage in make-believe play are happier, as measured by their expressions of interest and curiosity and by their smiling faces.

Images of fantasy and reality

Television drama, with its endless profusion of make-believe scenarios, has implications not only for the child's developing imagination but also for his or her ideas and expectations of the real world. Television drama may offer a rich source of material for make-believe, richer than the child has in his or her immediate environment. For those children whose parents are absent or too busy or highly stressed through illness or poverty, television may seem a dependable source of make-believe ideas. The casual conversation of many children suggests that their make-believe friends and fantasy figures are drawn from television characters, such as Bionic Man, Bionic Woman, Superman, or even the Cookie Monster.

Just as some of what the child sees on television is incorporated in play and fantasy, much of it affects his or her beliefs and expectations of what the world is like. Both effects may occur at the same time. For example, a child watching a dramatic program about a private investigator may imagine leading that kind of exciting, active, dangerous life. Basing the fantasy about such a life on the incorrect information gained from that program and other, similar ones, the child may develop inaccurate conceptions of the reality of an investigator's career. Since the conception of reality is inaccurate, the fantasy based on that conception is affected. Subsequently, the child may have the difficult task of distinguishing between his or her fantasy, his or her belief in reality portrayed on television, and the reality which later experience will eventually reveal. At the least, the eventual discovery that real life is not what television has made it appear to be can be disappointing and even disillusioning. Because some fantasies, especially those which enhance self-esteem or gratify wishes for adequacy and competence are not easily relinquished, even in the face of the harsh evidence of reality, some individuals will prefer to ignore or distort that reality rather than forsake the illusion on which the fantasy is based.

When fiction and reality merge

When dramatic presentations blur the line between fiction and reality and are also stimulating, exciting, or frightening, the young child may experience considerable anxiety or nightmares. In part, the fear stems from a dread about whether what happened in television might also happen to him or her. The child is confused between what is possible and what is only make-believe.

As anyone who has had a nightmare will attest, not being able to distinguish what is real and what is not can be extremely threatening. While such experiences are developmentally expectable, the individual qualities of children, and the parents' abilities to help, may make children more or less vulnerable to damage. Programs which confuse fiction with reality may hold the interest of a given young child precisely because they relate in a new way to conflicts with which he or she is concerned. But the programs will not help to resolve the conflicts as could his or her own fantasies. A parent, moreover, may be unavailable or unable to help. In such a case, the child's conflicted feelings may intensify, and anxiety will increase.

Various reactions to *The Wizard of Oz,* regularly shown on prime time television, illustrate the interaction between the evocative dramatic material and the unconscious preoccupations of the viewers, as well as the varying, often idiosyncratic nature of these reactions:

> For one boy, the greatest distress was watching the Tin Man rust into immobility. This boy had, from his childhood on, been extremely athletic; mobility was central to his psychic organization

and a vital ingredient to continued psychological stability. Another child with cerebral palsy could not move her right side. She was distressed by the easy magic which could give courage to the lion and brains to the scarecrow but could not cure her. Other children found the transition from a "real life" character to a witch, or watching the trees grab Dorothy and her friends, to be most disturbing.

Such a film provokes disturbance because it arouses anxiety around significant personal issues at a time when the ability to differentiate fantasy from reality is relatively new and easily upset. In *The Wizard of Oz,* despite its undisputed salutary impact, the reassuring and organizing difference between fantasy and reality is blurred. While the blurring adds considerably to the dramatic effect, it has uncomfortable consequences for some children.

In the more recent creation, *The Six Million Dollar Man,* the distinction between fiction and reality is even more ambiguous. Here, no wizard cues the child into the "let's pretend" nature of the story. Rather, an invisibly mechanized human being looks and dresses like anyone on the street. When he goes into heroic action, he does not change in appearance. Superman and Captain Marvel have, of course, flown through comic strip space for years, but their superhero costumes signal a situation of make-believe. In the case of fantastic machines, they are either there or not; Tom Swift's Gravitron is clearly none other than a fantastic machine. *The Six Million Dollar Man* is somewhere in between these superheroes and fantastic machines. He has machines built in as part of his powers, but he looks like a human being. He does miraculous things but keeps on looking human. The cues to signal whether the story is real or not are blurred. Children watching programs of this sort often show greater difficulty in resolving the feelings and fears the programs arouse. Such programs do arouse fears among the viewers. Lyle and Hoffman found that television did intrude on the dreams of first graders, and that 40% of the first graders they studied admitted to being frightened by shows on television.[3] When a confusion between reality and fantasy is added to ordinary fear, the effect can be terrifying.

The ability to fantasize

The effects of television on internal fantasy will differ for different children. Some will borrow the prefabricated fantasies and pursue these in their daydreams. Some will not have the time or freedom to daydream at all. Some will use television stories to protect themselves from their own too scary fantasties. Some will develop inaccurate views of reality. Some will become confused and anxious over what is reality and what is fantasy. Some will ignore the television story, since they have the ability and inclination to create and pursue their own fantasies.

Giving the viewer so much real and fantastic material, television could conceivably satisfy all of a child's fantasy needs and induce an atrophy of his or her own imagination. Would there not seem to be some significant consequence to the repetitive viewing of prepackaged, make-believe scenarios, all active, exciting, and in vivid color? Could a close attachment to the television set not lead to a poverty of ideas and a dependence on television programming for new ones? These questions are difficult to answer in the short run, for such effects are probably the outcome of long-term exposure. The cause and effect relationships, moreover, interact with other relationships, such as those between television viewing and passivity or compliance, and between television programming and the capacity for logical thought or problem solving.

Despite these confounding analytic problems, we suspect that television deters the development of imaginative capacity insofar as it preempts time for spontaneous play. Experience has shown that children who cease watching television do play in ways clearly suggesting the use of an active imaginary world. Resuming their viewing, the children decrease this kind of play. Research findings also suggest that children who are light television viewers report significantly more imaginary playmates than those who are heavy viewers. Those with imaginary playmates also play more freely and have more highly developed language.[4] Nursery school teachers are

"We'd rather play games than watch TV."

Photo, courtesy of Bill Owens

reported to have observed that juvenile play today is far less imaginative and spontaneous than that of the pretelevision generation.

On the other hand, carefully designed programming can enhance a child's capacity for imaginative play. Singer and Singer randomly assigned children to one of four conditions: (1) exposure to a model who taught make-believe games; (2) exposure to the *Mister Rogers' Neighborhood* program with a mediator interpreting events and directing children to focus on elements in the film; (3) exposure to *Mister Rogers* but with no mediating influences; and (4) a control group which received no stimulation or leadership.[5] Those children who worked with the adult had the most imaginative play. The adult's encouragement, support, and help apparently increased the persistence and elaboration of their play. Other studies, discussed later, highlight the value of *Mister Rogers* alone. This study emphasizes the importance of parental involvement and support, even with carefully designed programs.

We close this section with a conjecture concerning television and fantasy. Fantasy is commonly used by healthy individuals to manage impulses which would, if expressed directly, be embarrassing or dangerous. People who cannot use fantasy as a stabilizing, action-deferring mechanism are most likely to express these forbidden feelings in direct actions, however dangerous those actions may be. Research has amply documented that individuals convicted of violent crime have a markedly restricted capacity for fantasy. In some people, overdependence on television could contribute to the restriction of that capacity.

If extensive viewing of television drama does decrease a child's capacity for fantasy, it may thereby contribute to an increase in impulsive violence at later ages. In the absence of scientific evidence, such a long-term relationship must remain speculative. But it does indicate how broad and how insidious may be the reach of television drama into the human psychology.

DOES TELEVISION PROMOTE PASSIVITY?

There is the widespread concern, expressed by some teachers, social commentators, parents, and many others, that television promotes intellectual and physical passivity among its viewers, especially the younger ones, and that television may even be creating a whole nation of passive, dependent characters. *Passivity* is a term that is ambiguous in the extreme. In our discussion, we assume that passivity may be either physical or psychological. Passivity may be manifest in both a limited responsiveness to environmental stimuli, including television, and in a reduced capacity to initiate activity.

Physical inactivity is commonly equated with mental inactivity in two ways. A quiet, passive mind, it is assumed, inhabits a quiet, passive body. Similarly, physical inactivity is believed to produce a quiet mind. Neither assumption is true. In fact, the relationship appears to be inverse. Although the influence of the physical state on mental processes is neither simple nor clear, we do know that physically inhibited people or those with marked physical restriction often have hyperactive imaginations and well developed capacities for mental imagery.

People often think that since watching television requires rapt attention in a quiet posture, chronic viewing causes passivity. We find evidence to support both sides of this belief. In favor of this common assumption are the findings from a study reported from the University of Southern California.[6] A group of 250 elementary students judged mentally gifted were exposed to three weeks of intensive television viewing. Tests conducted before and after the experiment found a marked drop in all forms of creative abilities except verbal skills. The students had, in effect, become more intellectually passive.

Data from an unexpected source shed additional light on this finding. Attaching electrodes to the heads of children and adults as they watched television, Mulholland and Crown found to their surprise that the brain waves generated while watching exciting shows were those of low attention states.[7] People's output of alpha rhythms increased, indicating they were in a passive state, as if they were just sitting in the dark. The inference may be that television of any sort is a training course in the art of inattention. But not enough is yet known about such findings to permit solid conclusions.

We ought, moreover, to consider the evidence on the other side of the passivity question. According to some research findings, television watching itself may not be the immobilizing experience that some have alleged it is. Siegel has reported one study in which a camera, mounted above the screen to record the activities of the viewers, showed that only those under the age of six were quiet.[8] Adolescent and adult viewers did their watching while simultaneously making phone calls, ironing clothes, washing dishes, horsing around, drinking beer, eating junk food, and even talking with one another. School-age children watch with frequent interruptions to leave the room, move to another chair, or argue with a sibling. The high level of interaction among those watching in the same room casts some doubt on popular claims that television has brought family communication to a halt.

Relating passivity to viewing patterns, Himmelweit and colleagues found little difference in passivity between children described as viewers and those described as nonviewers.[9] They did find that very heavy, "addicted" viewers appeared to be more passive. Their interpretation was that while television may increase the opportunity for and the ease of withdrawal, the initiating causes of the passivity are environmental and personality factors. Heavy television watchers appear to watch because they are basically passive rather than *vice versa*.

Murray demonstrated that among six-year-old inner-city boys, heavy viewers (more than 33 hours a week) were interpersonally passive children who were more likely to forget instructions, to drift from one activity to another, and to prefer solitary play.[10] Casual viewers (less than 18 hours a week) were found to be less distractible, less likely to be bashful about playing with others, and more likely to initiate interpersonal contact. Since the passive children had shown the same traits at age three, he inferred that passivity leads to heavy viewing rather than the reverse. As Murray suggests, the "electronic peer is a more accessible playmate for a passive child"[11] and thus serves to reinforce his or her withdrawal. Schramm and his colleagues seem to agree in general with these findings. They feel that for most children television is not a primary cause of passive behavior.[12] Just as we observed in the case of effects on fantasy, the issue of passivity and its relationship to television viewing is most likely a complex interaction between the child's personality and his or her environment, on the one hand, and the use and content of television, on the other.

Concerning passivity, we should like to emphasize here that preoccupation with passivity as an undesirable state of being sometimes reflects strong cultural judgments about the importance of activity as a statement of strength, competence, and masculinity. Concerns about passivity also reflect a related apprehension that passivity probably leads to dependency, again a negatively valued state. Both the stigma on passivity and the fear of dependency are limited ways of responding to human conditions. In each case, while we may have concern, we think it is preferable to evaluate the real impact on the person in his or her world, not to place a value judgment on the condition itself.

DOES TELEVISION FOSTER DEPENDENT PERSONALITIES?

Psychiatrists often describe dependent personalities as those who expect nurture, support, and gratification in exchange for little or no action on their parts. We note that these expectations develop naturally in childhood in relationship to the mother. In earliest childhood, such expectations are developmentally appropriate. As the child grows, however, the expectation of unconditional nurture, support, and gratification is gradually replaced by increasingly more mature reliance on personal responsibility and initiative. If this process of maturation is interrupted or retarded, as can ensue from either deprivation or excess gratification, a dependent personality may persist long past the appropriate age and may markedly affect the individual's relationships with others and with his or her environment.

Because the television is always available and ready to stimulate, to advise, and to give comfort, and never requires independent action or responsibility on the part of the viewer, persisting dependent needs in many young children and some adults as well may be partially filled by the offerings of television. In such a way, television may serve a role as a "substitute mother" for those with felt needs for more maternal attention than may be available, helping to reduce their sense of internal emptiness.

The likelihood that television might serve many children as a security blanket seems even greater when one considers the early age at which children begin to watch. A 10-year-old child with very busy, frequently absent parents and with few satisfying relationships with other children spoke of his dependence on television. "TV is my best friend!" he asserted. Taking away his television privileges, he added, would be "the worst thing that could happen to me." It seems probable that this orientation toward television as an illusory companion may continue into adulthood.

Glynn and colleagues observed psychotic children receiving nurture and attention from their use of television:

> They all demonstrate quite clearly the special set of needs television satisfied—needs centering around the wish for someone to care, to nurse, to give comfort and solace... These infantile longings can be satisfied only symbolically and how readily the television set fills in. Warmth, constancy, availability, a steady giving *without ever a de-*

mand for return, the encouragement to complete passive surrender and development—all this and active fantasy besides. Watching these [viewers], one is deeply impressed by their acting out with the television set their unconscious longings to be infants in their mothers' [undemanding] lap.[13]

The researchers extend these findings to the normal child growing up and draw hypotheses with implications for personality development. They conclude that this relationship with the television results in increased dependent personality traits in some viewers, but that for most, significant interference with active, effective social adaptation does not occur. Those it affects most are individuals who find too little in the real world to sustain them without effort on their part.

Some contemporary television programming content obviously encourages the wish for passive gratification in everyone and may thereby create problems for some, especially the young. Some programs reinforce a "gimme" ethic, an expectation of comfort, reward, and benefit for little or no return. This something-for-nothing psychology is particularly evident in prize, game, and quiz shows, as well as in commercials and dramas. Such programs imply that great gains can be expected from very little effort. The prevalent consumption ethic, part of the normal developmental experience of the five- to seven-year old child, may be reinforced and extended beyond its usual period by the repeated programmatic and commercial messages that one can, in fact, get something for nothing.

Stemming from such an attitude is a low tolerance for frustration. When tasks do not respond in the same instantaneous way that they do on television, children may give up, failing to learn to enjoy independent activities that could give them more stable satisfaction. "They want everything to be easy—like watching the tube," lamented one first grade teacher.

Children can also come to expect instant gratification from other people. Teachers have suggested that even the acclaimed *Sesame*

Street, Electric Company, and *Zoom* have created problems. These shows continuously gratify the children through entertainment in order to hold their attention. "Kids today are accustomed to learning through gimmicks," wrote a Connecticut teacher, but "I can't turn my body into shapes or flashlights."

The reinforcement of infantile patterns at the expense of more mature ones is of concern to professionals who daily strive to help their patients free themselves from such regressions and fixations.

HOW DOES TELEVISION RELATE TO LONELINESS?

Television viewing apparently does affect social interaction and activity, but findings vary with the researcher and the age group studied. Lyle and Hoffman, for example, have reported that children who are high users of both books and television were also active in sports, recreation, and in other social activities.[14] Among nonreaders, however, high television viewing was associated with decreased social activity. These researchers also found that the first graders who reported the highest levels of loneliness were among the heaviest viewers and the most isolated children for their age. Television appears to reduce the awareness of loneliness for these children, but at the same time it lessens the likelihood of human contact. In a study of sixth and tenth graders, the authors found children were likely to watch television when they desired entertainment, relaxation, or relief from loneliness. Such relief can be a positive consequence, of course, as with a bedridden child, but it should be only a stopgap measure. Children, especially those who are bedridden or otherwise confined, need direct human interaction.

In many homes the television set is continually on, providing ongoing sound and image without regard to content.[15] It seems to fill

The constant companion

Photo, courtesy of Bill Owens

space for people who feel empty, depressed, or lonely. The continuous sound creates an illusion that others are present. Like the transistor radio held tight to the ear by the mid-adolescent wherever he goes, whatever he does, the constant television may be acting to provide a sense of contact and of involvement or to distract from the pressures and anxieties of inner stress.

For many people, silence is not a desirable state. Sounds are an inescapable and even desirable part of life, especially for many city people. Peace for many people seems to lie not in escaping from sound but in making sounds acceptable, comfortable, and reassuring, even though the sounds may then also be noncommunicative.

Television may accomplish this function. Television is often on even though every member of the family may be doing something else—playing cards, reading, studying, cooking in another room, even using the vacuum cleaner. "It helps me concentrate," says the student. "It gives me a feeling of life around me," says the housewife, "something going on—friends at hand." "It's sort of scary without it," said one school-age child.[16]

Used repeatedly and habitually, television may be a way to avoid solitude or silence. A person who uses it this way may fail to learn to spend time alone and to use that time for intellectual, emotional, or creative activity. For some children and adults television acts in specific competition to the often painful experience of self-awareness, suppressing the intrusion of appropriate self-critical thoughts or feelings, along with more obsessive worries. Because of its special properties, dramatic programming, in particular, may also inhibit creativity. If one is never alone, the experience of having daydreams, fantasy, unguided thought, and creative activity is virtually eliminated.

While it can reduce loneliness, television also appears to reduce the impetus to learn how to be alone. In effect, it encourages a person to escape from solitude. And, of course, the more the television itself is the solution to loneliness, the less likely it is that such people will develop relationships with other human beings.

REFERENCES

1. Bettleheim, B. THE USES OF ENCHANTMENT. New York: Random House, 1975.
2. Singer, J.L. "Television, Imaginative Play and Cognitive Development: Some Problems and Possibilities." Invited address to the Division of Educational Psychology, American Psychological Association, San Francisco, Calif., 1977.
3. Lyle, J., and Hoffman, H.R. "Children's Use of Television and Other Media." In E.A. Rubinstein, G.A. Comstock, and J.P. Murray (Eds.), TELEVISION AND SOCIAL BEHAVIOR, Vol. 4. Washington, D.C.: Government Printing Office, 1972, pp. 257-273.
4. Caldeira, J., Singer, J.L., and Singer, D.G. "Imaginary Playmates: Some Relationships to Preschoolers' Television Viewing, Language and Play." Paper presented at Eastern Psychological Association, Washington, D.C., March 1978.
5. Singer, J.L., and Singer, D.G. "Can TV Stimulate Imaginative Play?" *Journal of Communication,* 1976, *26,* 74-80.
6. "What TV Does to Kids." *Newsweek,* 21 February 1977, p. 68.
7. "Learning to Live with TV." *Time,* 28 May 1979, p. 49.
8. Siegel, Alberta E. Personal Communication, 1979.
9. Himmelweit, H.T., Oppenheim, A.N., and Vince, P. TELEVISION AND THE CHILD. London, Oxford University Press, 1958.
10. Murray, J.P. "Television in Inner-City Homes: Viewing Behavior of Young Boys." In E.A. Rubinstein, G.A. Comstock, and J.P. Murray (Eds.), TELEVISION AND SOCIAL BEHAVIOR, Vol. 4. Washington, D.C.: Government Printing Office, 1972, pp. 345-394.
11. Murray, J.P. "Television in Inner-City Homes: Viewing Behavior of Young Boys." In E.A. Rubinstein, G.A. Comstock, and J.P. Murray (Eds.), TELEVISION AND SOCIAL BEHAVIOR, Vol. 4. Washington, D.C.: Government Printing Office, 1972, p. 365.
12. Schramm, W., et al. TELEVISION IN THE LIVES OF OUR CHILDREN. Stanford, Calif.: Stanford University Press, 1961.
13. Schramm, W., et al. TELEVISION IN THE LIVES OF OUR CHILDREN. Stanford, Calif.: Stanford University Press, 1961, p. 158.
14. Lyle, J., and Hoffman, H.R. "Children's Use of Television and Other Media." In E.A. Rubinstein, G.A. Comstock, and J.P.

Murray (Eds.), TELEVISION AND SOCIAL BEHAVIOR, Vol. 4. Washington, D.C. Government Printing Office, 1972, pp. 257-273.
15. Medrich, Elliott. "Constant Television: A Background to Daily Life." *Journal of Communication,* 1979, *29*(3), pp. 171-176.
16. Menninger, Karl A. Television: "The Comforting Presence." (With Jean Menninger) *TV Guide,* 18 May 1968, p. 7.

4

THE PROGRAMS: DRAMAS WITH LIVE ACTORS

The programs in which we psychiatrists are most interested are the television dramas which children watch. These dramas are numerous and varied. The review which follows touches on their major properties, reviews the research, and expresses our own opinions concerning three types of dramatic programming—dramas with live actors, cartoons, and dramatic educational programs for children.

Television drama is, in our view, popular because it offers the viewer an opportunity to participate vicariously in imagining situations which are exciting, frightening, or dangerous, embarrassing or tempting. Because the role is a vicarious one, the viewer has no responsibility for either creating the situations or resolving them. He or she is free to enjoy the fantasy without guilt or obligation.

The typical sequence of the television drama includes suspense, climax, and resolution. Combining both visual and auditory stimulation, television drama exposes the relatively passive viewer to many intensely arousing experiences. The dramatic content may be further enhanced by the use of such devices as depiction of violence, quick cuts, rapid action, and mood music. Relaxed and sitting in a comfortable chair, the viewer can simultaneously experience the emotional roller coaster of dramatized television and remain completely safe. Removed from the danger presented and yet able to participate, the viewer is protected by the audience role and the comforting knowledge that the program and the hero will return next week, same time, same station. This security encourages the magical fantasy of having faced great danger, even death, and having survived, and permits many viewers to offset pervasive feelings of helplessness in a complex world. For a few fleeting moments, they can achieve a sense of personal competence and even heroism.

In moderate doses, such experiences are unlikely to be harmful. Indeed, they can be supportive and healthy for many people. Moderately used, escapism, fantasy, and vicarious experience with dangerous and exciting events are characteristic of a healthy, mature personality. However, some people, particularly young children, may be vulnerable to excessive use of television. Continuous exposure may exaggerate their potentially self-destructive patterns of escapism or encourage them to copy problem-solving methods which are of only limited usefulness in the real world.

We have some particular concerns with the content of the television dramas enacted by live players. The stories and their plots, as we shall suggest, inaccurately portray the origin and place of conflict in human relations and how conflict is resolved. A preference for the quick and simple solution prevails, and violence is frequently the means. The characters, too, are often simplified and unreal. The result, as the following discussion suggests, is that children who view these dramas over a period of time may develop highly unrealistic ideas about people, problems, conflict, and violence.

THE PLOTS

Conflict and conflict resolution

Conflict is central to any drama. In the typical television scenario, it may take the form of a struggle between individuals or groups for something of value, or an effort to clear someone of suspicion or guilt or to solve a crime. Usually the conflict illustrated on television is interpersonal. Seldom is it internal; conflicts between one's conscience and one's urges or temptations are rarely portrayed. Thus, the usual television scenario grossly externalizes and thereby oversimplifies the complex nature of human conflict. The result is a psychological world of spurious clarity and simplicity, quite unlike the real one in which personalities and interactions are complex, motives unclear, and outcomes ambiguous.

The conflicts of daily life are rarely simple choices between good and evil or right and wrong. Some involve desirable ends that have different costs along different dimensions—short-term versus long-term costs, time costs versus money costs. Some involve moral issues, such as a conflict between self-gratification and an obligation to help someone else. Some involve the difference between what one wants to do and what one ought to do.

Daily conflicts that appear to be largely interpersonal ones may be complicated by significant intrapsychic issues. An individual may, for example, project attitudes or ideas which are consciously distasteful but unconsciously attractive onto other people, who are then condemned, criticized, attacked, or rejected. It is often this contribution of one's inner dilemmas to conflict with another person which exaggerates the disagreement and makes its resolution difficult.

Television programming for the most part ignores this common human response to conflict. Instead, conflict is simplified first by reducing it to dichotomies which are easily recognized and understood: good/bad, right/wrong, win/lose, yield/resist. Then these opposing positions are assigned to separate individuals in the story as their primary role characteristic. Splitting and displacing the two sides of a conflict in this way does keep the story line clear and understandable, but at the cost of obscuring the fact that *both* sides of a conflict can usually be found within each party.

In the same manner, personifying good and evil through the characters of the hero and the villain is a time-honored theatrical device. Not uncommonly, the characters of televised drama, particularly the heroes, do depart from the idealized models of behavior and rely upon violence, dishonesty, or other methods characteristic of a villain. But rather than despoiling the image of the hero, the use of such methods by the hero legitimizes them. May has described the phenomenon of "contrast conceptions"; once the "good guy" has been identified, everything he does automatically becomes good, and *vice versa* for the "bad guy."[1]

In an earlier program series entitled *Switch*, the "good guys" explicitly employed deceit and dishonesty as primary tools. The fact that both protagonists were likable, while their targets were not, and the fact that their methods seldom employed violence tended to minimize the negative qualities of the essentially immoral methods they used. In any case, the crimes they fought seemed to justify almost any treachery, especially if it could be performed in a humorous and imaginative way. But what was the covert message? That any dishonest or even illegal act is all right if evil is the target? That a virtuous end justifies use of an evil means? Either one or both messages clearly come through, and because the characters are likeable and the action warmly humorous, the viewer is inclined to affectionate agreement.

Conflict resolution. Most real-life conflicts do not have neat, clean endings, and they are seldom resolved rapidly. Indeed, many are never fully resolved at all. There are often unexpected consequences or costs that were not anticipated and residual feelings of doubt ("was I right?"), uncertainty, or even guilt. Moreover, ordinary conflicts often do not stay settled; they reappear in a different form or with different persons; or, as expressions of significant internal conflict, the individual may even recreate them. Further, when real-life conflicts are resolved, it is usually by compromise and seldom in the all-or-nothing way of television.

Television commonly treats conflict resolution as a zero-sum game in which everything gained by one side is lost by the other. The process is often organized like a competitive game. "Evil" must be defeated simply to prevent serious loss to the representatives of "good." Only after several narrow escapes, near defeats, and tight squeaks does "good" come through. When "good" wins, it wins all, and "bad" is annihilated—often literally. In show after show, rewards and punishments come quickly, uncontaminated with regret, sadness, or any acknowledgment that there is some cost for every benefit. Crises are completely resolved, unambiguously and often violently. Authority invariably triumphs, often by using more violence than has been employed by the evil ones. This winner-take-all concept is very comfortable for the viewer. Since the loser is the epitome of evil, his defeat produces only good feelings, no matter what the cost. Indeed, the higher the cost of vanquishing the loser, the stronger the feeling that his annihilation is just what he deserved.

Although all narrative art involves the condensation of experience in time, television drama seems to exaggerate this process by conveying an unrealistic sense of what can be done in any particular amount of time. The problem is not just that the rapid move from the statement and elaboration to the resolution of the problem occurs in 30 or 60 minutes (minus the time for commercials); a program commonly has scenes covering a period of hours, days or months. The problem is that such resolution, by developing quickly and completely, conveys an implication that speedy management of a problem is not only possible, but altogether usual and desirable. The daily experience of most people fails to confirm this model.

Real-life conflicts seem to take a long time to develop, and even more time and effort to resolve. Yet Dr. Welby, with a few well-chosen phrases, can produce instant change of attitude and behavior in a defiant teenager. Doing so, he reinforces the myth of easy, immediate resolution that fits the expectations for instant gratification that are already so well established in our culture.

As well as giving the unreal impression that life's conflicts are resolved with simplicity and speed, the televised portrayal of conflict resolution generally relies heavily on action. Certainly the visual nature of the medium lends itself to nonverbal expression (chiefly action), and this capability is unquestionably one of television's strengths. Because action is an occasionally effective and appropriate method of conflict resolution in contemporary society, it is natural that it be used in this visual medium. But action is also expedient from the point of view of the television writer and producer: it is easy, effective, and fast, and it gets more attention than talk.

Constant reliance on action to solve problems unfortunately implies that action is the only method of problem solving that works. Dramatized television episodes in prime time are hardly ever resolved through talk, argument, or debate. Negotiation and compromise are almost never used, except fraudulently, for instance to entrap a criminal holding a hostage. The exclusive commitment to action suggests that talk is needless, and perhaps useless, and that it is largely resorted to by those who do not know what to do. This implied association of action with strength and talk with weakness is well illustrated in a bit of dialogue: "Talk is cheap. My Colt .45 does my talking for me. No real man would waste his time talking like this. Let's move!" This and a thousand other utterances from television scripts testify to the high value placed on action and to the contempt for talk. An exclusive action model for resolving differences is of very limited application in the complex daily lives of most viewers. Based on action (and especially violent action), the conflict resolution in television drama offers little useful guidance to the young viewer. Its seductive simplicity may well be dangerous.

In families in which violent action is commonly used to resolve conflict, the violent action they see on television is familiar and acceptable and reconfirms the appropriateness and legitimacy of violence to solve problems in daily living. Such reinforcement may discourage parent and child from working with less familiar nonviolent and verbal modes of communication to resolve conflict. Action-oriented television drama may, thus, reinforce what is actually a primitive, impulsive, action-without-thought style that characterizes many adolescents and young adults.

In much television programming, the violent climax and resolution occur at the very end of the program, allowing no time or opportunity to show the effects of the violent resolution. Resolution appears not only swift and active but also effortless and without consequences. The expectation of rapid and complete resolution, together with an intolerance for anxiety, discomfort, or delay, is an essentially infantile perspective, characteristic of immature personalities and very young children. Yet it is this attitude that television offers and reinforces again and again in its idealized portrayal of conflict resolution. Although little scientific, verified evidence is available to corroborate such a conclusion, we presume that the repeated presentation of behavior which endorses such an infantile perspective cannot be favorable, particularly for children. The persistent viewing of resolutions which are quick, neat, complete, and unambiguous builds the unrealistic view that immediate solutions to problems are the rule and that the exceptions one meets in daily living are due to someone's failure, inadequacy, or even malevolence. A child growing up with such expectations is headed for harsh disappointments.

The role of violence

Prevalence and significance. Violence has been a part of television entertainment drama since the medium itself began. As early as 1952 and 1953 Smythe reported finding some 3,421 threats and actual acts of violence in one week on seven New York City channels, an average of 6.2 threats or acts per hour.[2] He also indicated that television directed at children had more than three times this average. Twenty-five years later Comstock drew the same two major conclusions—that there is a great deal of violence on television drama, and that violence is very frequent in television directed at children.[3] Other researchers, particularly Gerbner, have made it clear that the frequency of violence in television programming has not materially diminished since his surveys began, in spite of the introduction of the "family hour" concept in 1976.[4] During the 1979 season, 70% of all prime time programs still contained violence. Nearly 54% of all leading characters were involved in some violence, about the same as the year before. In weekend daytime (children's) programs, 92% of all programs contained some violence, down from 98% in 1978. The rate of violent episodes was 17 per hour, down from 25 the year before.[5]

Gerbner and his colleagues have defined violence as "the overt expression of physical force (with or without a weapon, against self or others) compelling action against one's will on pain of being hurt and/or killed or threatened to be so victimized as part of the plot."[6] Based on this definition, they have developed a Violence Index, which they apply annually to television programming. Television industry representatives have criticized the measure for being too arbitrary and for counting comedy accidents and acts of nature as part of a normative count of violence. Thus, it may be overinterpreted by people who misunderstand it.[7] Nonetheless, the Violence Index and Gerbner's careful application of it represent a systematic attempt to discern patterns of violence in television and to identify trends deserving special attention.

In 1977 Gerbner reported:

> ...television violence increased sharply [over levels reported the previous year] in all categories including "family viewing" and children's program time on all three [commercial] networks. The increase [they reported] resulted in the highest Violence Index on record. The only score that [came] close to the current record of 203.6 was the score of 198.7 in 1967, the year of turmoil that led to the establishment of the Eisenhower Violence Commission....[8]

Television is, of course, not the only medium to promote violence. Violence is and has forever been a part of comics, paperback books, and the movies. The movies' output of violence and horror has, in fact, been proportionately far higher than television's. It is also true that traditional art forms, including the theater, have always been preoccupied with violence, war, and mayhem. But the historical portrayal of violence in the earlier media differs significantly from the television portrayal and medium. Somers describes the difference:

> Traditionally, violence was portrayed in a context of high tragedy (the Bible, Greek and Shakespearean tragedy, Tolstoy's *War and Peace*, etc.), fantasy (fairy tales, Wagnerian legends, early cowboy and Indian stories, etc.), or outright slapstick (Marx brothers, Charlie Chaplin, etc.). It was generally not related to contemporary people or current "real life" situations. By contrast, much of to-

day's movie and television violence is presented in the context of ordinary life and routine problem-solving. The implications for the viewer are very different: his personal identification either with the murderer, the murdered or both, is likely to be much greater.[9]

The violent movie or theatrical performance undoubtedly had an impact. But the intermittent and limited accessibility of the theater and the movies contrasts sharply with the very high accessibility and constant availability of violent television programs for contemporary viewers, particularly children. Moreover, the movies and the theater did not have to plan for a violent cliff-hanger every eight minutes in order to hold the audience through the next commercial.

Effects on viewer aggression. The question of causal relationships between televised violence and increased aggressiveness in viewers has attracted much time, attention, and energy. Between 1952 and 1974 Congress held seven hearings on the subject. A National Commission on the Causes and Prevention of Violence worked for several years and published a major report on the subject.[10] In 1972 the Surgeon General's Scientific Advisory Committee on Television and Social Behavior published a five-volume report.

Bandura's famous Bobo Doll studies in the early 1960's demonstrated that people could learn aggressive behavior from observing others and that television therefore could be responsible for encouraging violent behavior.[11] Bandura's findings stimulated a great number of studies which largely focused on demonstrating the power of television to provoke imitative aggressive behavior.

Imitative violence, instances in which violent acts have been copied from television drama and acted out on random victims, is infrequent. Such violence is, however, so dramatic that instances of it typically capture national attention.

> In October 1973, six youths in Boston set upon a young woman carrying a can of gasoline to her car, forced her to douse herself with the gasoline and then set her afire, burning her to death. This

> happened two days after the nationwide showing of a movie, *Fuzz*, a police drama set in Boston, which contained a scene portraying teenagers burning a derelict to death "for kicks."

In several instances the families of the victims of such violence have sued the networks and producers.

> In 1978 NBC was unsuccessfully sued on behalf of a young girl who had been violently raped on a San Francisco beach by other girls after the assailants had seen an almost identical scene in a movie, *Born Innocent*, broadcast by NBC shortly before.

In spite of the implication of causality which imitation suggests, research has in fact not demonstrated a simple one-to-one connection between televised violence and violent behavior. Liebert, Neale, and Davidson, for example, were unable to conclude that the level of criminal violence is influenced by television violence.[12] Imitative violence, like other human behavior, is very likely the result of a complex interaction between the personality, the environment, and the particular circumstances. As such, imitative violence may be an expression of individual or group needs. Boredom, for instance, may lead a person or group to indulge in violence: it produces excitement and serves to test courage and strength, or provide social acceptance. Individual preoccupations with authority, fears of inadequacy, and participation in activities designed to prove group dominance may be considerably more important influences on the person's violent behavior than the televised example the person later imitates. Because of such influences, young people are sometimes especially vulnerable to televised violence.

Giving vulnerable young people examples of criminal pursuits may encourage them to express themselves and to release their unusual tensions in antisocial ways. Indeed, a study of 100 juvenile offenders commissioned by ABC found that no fewer than 22 confessed to having copied criminal techniques from television.[13]

Still, the role of television is ambiguous. Is it a teacher, a facilitator of impulses already in the viewer, or does it implant and coax forth new impulses into the open? Wertham, long concerned about the deleterious effects of media violence on young people, concluded that no matter how it does so, television blunts emotional and moral responses to both real and televised violence and also serves as a uniquely effective "school for violence":

> Whether crime and violence programs arouse a lust for violence, reinforce it when it is present, show a way to carry it out, teach the best method to get away with it or merely blunt the child's (and adult's) awareness of its wrongness, television has become a school for violence. In this school young people are never, literally never, taught that violence is in itself reprehensible. The lesson they do get is that violence is the great adventure and a sure solution, and he who is best at it wins.[14]

Research findings. More than 4,500 studies of televised violence had been done by 1978, but they varied greatly as to the methods they used, their theoretical perspectives, the issues they examined, and the nature of their findings. Some studies have used survey instruments to identify correlations regarding the viewer, his or her behavior, and the material viewed. Other studies used experimental methods to examine the behavior children learned from television after several weeks or months following their viewing. Other experimental studies have tested subjects immediately before and after viewing samples of television programming to determine program effects on viewers' behavior, including antisocial and positive social behavior. These studies seldom utilized knowledge of child developmental phases or clinical understanding, focusing instead on definable parameters in the laboratory situation and subsequent observable behaviors. From the perspective of the clinician, an understanding of the impact of such events on an individual is heavily dependent upon the meaning of those events to that individual.

No account is taken by many of the researchers whose work dominates the field, and many of whom we quote, of the probability that individual reactions to televised violence are primed by predisposing aspects of that indivual's personality and affected by important interpersonal variables such as relationships to important others. However, because personality variables are crucial to the understanding of the impact of television viewing on specific individuals and particularly the cumulative effects of such viewing, studies which omit consideration of personality variables are not likely to yield clear and unequivocal answers to questions about the effects of television on different sorts of individual personalities.

Dramatically illustrating the significance of the meaning of a particular television drama for the individual viewer, Redl described the distressed reaction of a group of boys at a residential treatment center after viewing a movie on television:

> The movie depicted a loving, caring middle-class family. The scene that was so upsetting showed an eight-year-old boy, secretively leaving his house late at night while his parents were sleeping, to sleep in a tent in the back yard. Several hours later the boy is drenched by a severe thunderstorm. He is shown frightened and dripping wet, running to the back door, only to find that he has locked himself out. His parents, responding to his frantic knocking, come running downstairs, and the moment of confrontation arrives.
>
> It is at this moment that expectations of outcome become so important. The boys at the residential treatment center expected the child in the film to be severely reprimanded and possibly punished physically. When this did not occur, they became extremely anxious. In fact, the parents responded with concern, not anger; comfort, not rejection; understanding, not punishment. He was embraced, wrapped in a warm towel, and comforted before any discussion occurred about his reasons for being outside so late without permission. In the end, he was given nothing more than a mild reprimand and kindly parental advice regarding such experimentation without first discussing it with his parents.
>
> To see parents respond in this unfamiliar way was too much for

many of these youngsters. The movie scenario enhanced awareness of their own unfulfilled needs, with feelings of anxiety, anger, depression, and loneliness. As a result, a chaotic night ensued with much fighting, as these youngsters struggled to handle the many conflicting and confusing feelings stimulated by the movie. A scene which would generally be thought to be positive and comforting was, for this specific population, quite the opposite.[15]

Extensive studies have illuminated some of the many other intervening variables which determine whether, how much, and what kind of aggressive behavior might be expected. Dorr outlined the categories of factors which must be taken into account.[16] These factors include the type of program in which violence is presented and the clarity of the message that the program presents about violence. Another complex of factors stems from the extent to which a given program encourages a person to learn about various behaviors, about the situations in which those behaviors may be acceptable or useful, and about how to perform actions similar to those seen on television. Other relevant factors include the viewer's age, sex, and previous history of aggressive behavior, and the time, place, and circumstances of viewing.

Using these and other variables, some research has expanded the limited focus on correlational patterns to recognize that effects may take the form of gradual and generalized psychological changes. Among these effects is the "disinhibition" or loosening of the personal constraints on aggressive behavior in general. Liebert noted that, among the five correlational studies published in the Surgeon General's Report, all, regardless of their method, found some association between viewing television and generalized aggressiveness.[17] He also called attention to the impressive findings of Lefkowitz's longitudinal study of 427 children over a 10-year time span. This study demonstrated that *"the greater a boy's* [but not a girl's] *preference for violent television* at age 8, *the greater was his aggressiveness* both at that time and *ten years later."*[18] Reexamination of the data by others has continued to support the inference that there was a causal relationship.

The Programs: Dramas with Live Actors 65

The findings have been reconfirmed in a study supported by the Columbia Broadcasting System and conducted in England by William Belson.[19] Examining the exposure to violence and the subsequent social behavior of a group of 1,565 boys between the ages of 13 and 16, Belson found that those boys who watch a good deal of television violence between ages six and nine are nearly twice as likely to indulge in violence during adolescence as boys who watch much less.

Liebert cited several studies in which children were randomly assigned to groups varying in the amount of exposure to televised violence, and noted that such experiments have provided further evidence for a causal relationship via "disinhibition," as well as generalized effects reflected in a lowered sensitivity to aggression and decrements in cooperative behavior.[20]

Whatever their limitations from the clinician's point of view, the great number of studies of the impact of televised violence have generated several hypotheses for which there is empirical support.[21]

1. People, and especially children, *do* manifest greater aggressiveness after viewing televised violence. Program characteristics which appear to promote aggressive behavior after viewing are as follows:

    - live portrayals of violence, especially between people who have a close relationship;
    - a suggestion that the violence is justified, i.e., done on behalf of a good cause, done against a person depicted as bad or evil;
    - a suggestion that violence pays off or is rewarded and/or not punished;
    - a realistic depiction of violence, especially if sanitized, i.e., shown without evidence of the bloody, painful disabling consequences;
    - the presentation of violence in circumstances recognizably similar to those of the viewer, i.e., targets, implements, or cues similar to the real-life milieu;
    - the performance of apparently socially acceptable violence

by a hero with whom viewer identification is easy, i.e., the good guy, the representative of the law;
- frequently used violence depicted clearly as the favored method for attaining goals; and
- frequently repeated violence.

2. Aggressive behavior following viewing usually appears to reflect a breakdown of inhibitions against such behavior rather than being a simple modeling or imitative process.[22]
3. Exposure to portrayals of televised violence may desensitize people, particularly young people, to violence in their environment, retarding their response to it.[23]

Some other effects: Beyond the issue of causality, Gerbner and Gross have suggested that exposure to large amounts of televised violence may induce important secondary effects and convey unintended messages about the nature of law and authority in our society.[24]

The fact that so much of the violence on television drama is actually instigated by law enforcement agents (as "good guys") suggests the idea that only violence by criminals is bad. Violence in the service of law enforcement is acceptable and fully justifiable, especially if it is used against individuals who have been clearly labeled as evil or criminal. When the "establishment" uses violence, an individual can feel justified using it as well, if he sees himself fighting evil in the same way. Such a pattern is evident in an interesting study of motivation and aggression. Students witnessing a beating in which the victim was portrayed as deserving the punishment were observed to be more aggressive than those watching a beating in which the victim was portrayed as an unfortunate victim of circumstances.[25]

Viewing violence appears to affect young people's general view of the world and their attitudes toward others. From their observations of the differential effects of heavy and light television viewing on a large number of teenagers, Gerbner and Gross concluded:

The Programs: Drama with Live Actors

> ...[television]...dramatically demonstrates the power of authority in our society, and the risks involved in breaking society's rules [but] violence-filled programs show who gets away with what, and against whom. It teaches the role of victim and the acceptance of violence as a social reality we must learn to live with—or flee from.[26]

They found that adolescents who watch a great deal of television see the real world as more dangerous and frightening than those who watch very little. Moreover, heavy viewers are less trustful of their fellow citizens and more fearful of the real world. Some 35% of heavy viewers were more likely to agree that one "can't be too careful." When these same viewers were asked to estimate their own chances of being involved in violence during any given week, 59% *overestimated* the likelihood. They estimated that their chances were between one and five in 10, rather than the more accurate probability of one in 100. Only 39% of the light viewers made the same pessimistic miscalculation. Gerbner also noted that excessive fearfulness was especially striking among people under 30 who have been exposed to television since birth and have never known life without it. More recently, Gerbner and his colleagues offered further strong support for their theory that the saturation of television with violence cultivates mistrust, apprehension, danger, and a "mean world" perception among heavy viewers.[27]

Although recent findings of Doob and MacDonald have suggested that controlling for variations in crime frequency between neighborhoods eliminates the relationship between heavy viewing and excessively fearful attitudes, the Gerbner thesis is nonetheless provocative and worth further study.[28] This is particularly so since data about the psychological status of the viewers studied are frustratingly absent from both Gerbner's and Doob and MacDonald's work. Fearfulness in the viewer could result from a variety of factors other than television viewing or from a combination of viewing with these other factors. Significant parent-child tensions, family conflict, or even a preexisting fearful point of view about society could be underlying factors, reinforced, perhaps, by television viewing. How important such factors may be in relation to exposure to televised violence cannot be answered without further data.

Why violence? Those who justify the use of violence on television claim that life is inherently violent and that overt violence in drama is not only necessary to the story line, but is consistent with reality. One unidentified industry official argued that violence has always been a peculiar American fascination, implying that using it in televised drama is merely responding to this fascination. "Violence is the basis of all drama," he said. "Violent conflict is necessary to a story. Without violence, at least implied, there is no drama."[29] The ignorance of such spokesmen is as frightening as their bland indifference to the psychosocial implications of such attitudes. It is conflict, not violence, that is essential to drama.

The heavy dependence of televised dramatic entertainment on violence plainly reflects the fact that violence "works." Violence attracts and holds mass audiences, or so it is presumed by the producers. Recent evidence, however, suggests that violence does not in fact enhance popularity. Diener and DeFour found no correlation between the amount of violence in the episodes of 11 series over a season and their Nielsen ratings; they coded "violence" only in cases of actual physical or verbal abuse, not in cases of quasi-violent action elements, such as car chases.[30] They also found that the degree of perceived violence was unrelated to liking for a given episode.

Continued portrayal of so much violence in television drama may arise in part from a failure to distinguish between the superfluous and the essential, between violence and dramatic conflict which is in some other way exciting, arousing, or engaging. Aside from the previously noted general (but erroneous) assumption that conflict *equals* violence, there is a widespread erroneous notion that violence *equals* entertainment. The latter view, however, is contradicted by the results of an experimental field study of 183 couples watching television in the natural settings of their homes.[31] The study found that programs such as *The Waltons*, *Little House on the Prairie*, and *Apple's Way* were both more entertaining and significantly more emotionally arousing than violent programs such as *Hawaii Five-O* and *Streets of San Francisco*.

As a dramatic convenience, violence adds to the program's excitement and gets the viewer's attention. Moreover, as it portrays physical jeopardy in the course of a conflict, violence accentuates and clarifies the issue the story depicts. Easing viewers' perceptions is important in a medium that must make its point visually.

Violence as a dramatic convenience does more than get the viewer's attention. It portrays physical jeopardy in the course of a conflict and thereby accentuates and clarifies the issue the story depicts. Easy perception is important to a medium that must make its point visually.

Violence is undeniably exciting. The appeal of violence, however, goes deeper than just excitement. Violence is related to the vicarious expression of the aggression and hostility in all of us, our need to punish evil in ourselves and others, and the wish to deny our own vulnerability. Yet violence is a forbidden impulse except under certain sanctioned circumstances. Our culture reinforces the parental teaching that violence against members of the family must be curbed, except for that violence which parents use to control their children. Television offers us an opportunity to vent our prohibited violent impulses safely, albeit vicariously, on persons who are both disliked and negatively valued, and to do so without the pangs of conscience that normally ensure against our behaving violently in real life.

But the value of television as a route for the vicarious discharge of aggressive impulses is more than offset by its apparent tendency to stimulate aggressive behavior. We have noted the tendency of televised violence to loosen viewers' inhibitions against expressing aggression. But some observers have suggested that an awareness of the aftermath of aggression could help to inhibit aggression in the viewer of television violence.[32] Ironically, attempts to include such evidence of the consequences of violence, as part of news reports on the Vietnam War, for example, have been criticized as being insensitive, as glorifying violence, and so forth. The wish to experience the excitement of vicarious violence without having to acknowledge its painful reality is apparently very strong.

While televised violent episodes are themselves abundantly laced with noises and actions which emphasize the destructive quality of the event, seldom do we see the consequences of that violence. Viewers do not see the bloody, messy results, the pain, suffering, paralysis, and death which result. Their knowledge of violence is partial and distorted. Their understanding would be more complete if they customarily witnessed the weeping and anguish of the victims' families, the funerals, the sudden poverty, and the aloneness that are the aftermath of violence.

It might well be that a more honest depiction of violence would help to reduce the unalloyed pleasure it is capable of generating, as well as reinforcing some capacity for inhibition of violent action. At least the information delivered to viewers would be more accurate—an aspect important for children, who are prone to an uncritical acceptance of the information they receive from television that violence is safe, useful, and justifiable.

THE CHARACTERS

What the characters do in the course of a televised drama can become, and for some viewers does become, a basis for real-life action. How people deal with each other, how they seem to feel, and how they speak may become models for viewers to emulate. Exactly what the causal linkages are and how this process works are, as we have noted, complicated questions. The modeling may be done consciously, in an imitative manner, or it may happen unconsciously, over time, through the process of identification as a person, especially a young one, views programs repeatedly.

Of interest to psychiatrists are distortions in the presentations of human beings and the possible consequences from the viewer's perspective. We are especially concerned with the processes of imitation, identification, and modeling in children.

Stereotypes

In general, we find that dramatized television portrays people in such a simplified and inaccurate way that the characters are crude stereotypes. Occupational roles and workers, women, minorities, the aged, single parents, the handicapped, and various ethnic groups are quite commonly presented as stereotypes and given few individual characteristics. The development of central characters in the drama is quite limited, often intentionally so. The intent is to highlight only those characteristics necessary to the image of the hero or the villain. Secondary persons are bland and featureless, quite literally "supporting" and without dimensions of their own.

We are interested in these features of character portrayal and choose to discuss them in particular because we believe that character stereotypes and oversimplified (and violent) plots combine to promote dehumanization in the viewer. And this, particularly for children who view repeatedly, can be psychologically and socially damaging.

The work force. In its depiction of the world of work, television distorts reality by disproportionately emphasizing certain occupations and roles and ignoring a great many others. The labor force on television bears little resemblance to the real population of workers. DeFleur noted in 1964 that "professional workers were substantially overrepresented. Nearly a third of the labor force on television was engaged in professional occupations of relatively high social prestige."[33] Likewise a third was involved in enforcement or administration of the law. Occupations that were represented were represented stereotypically: lawyers were clever, actors were temperamental, and so forth.

Some nine years later in 1973 Seggar and Wheeler found the same pattern.[34] Professional and managerial roles were still overrepresented in the television picture of the work force. Occupations with low prestige were underrepresented except for the service area, and there minorities were represented more often than whites.

Men and women. Men play the lion's share of leading roles. Some time ago Head noted that there were twice as many men as women in TV character roles,[35] and DeFleur found that women were portrayed working only half as often (16%) as they work in reality (31%).[36] Since then the proportion of working women has steadily increased, but in 1973 working women were still grossly underrepresented on television. Only 18% of the television labor force was female.[37]

In a study of personality characteristics and nonoccupational roles, Tedesco found that men's roles were more likely to portray bad persons who were unsuccessful, unhappy, unmarried, violent, and serious as opposed to comic.[38] Women's roles, on the other hand, were more likely to be happy, married, nonviolent, and comic as opposed to serious. Tedesco also reported that the programs distributed personality traits to characters in a stereotypic way. Women were typically more attractive, sociable, warm, and peaceful; men were more powerful, smart, rational, and stable. In each case, the characteristics reinforce the stereotypic views of masculinity and femininity.

Central roles in dramatic television portray men who are mostly single, middle- and upper-class, white, American, and in their twenties and thirties. Seldom do they appear to have significant lives apart from their primary characterization; rarely do they worry about matters other than the central task of the drama of the moment. The many who are either single or without sharing relationships with others convey images of freewheeling, devil-may-care action. They have virtually no responsibilities or encumbrances, except those the dramatic tasks entail. The leading man is rarely burdened by real-life constraints, like a family. When present, the families are subordinate to the story line. They neither impose the irrelevant interjections, obstructions, competition, and confusion nor offer the support, assistance, and guidance that the typical family does. Such family shows as *The Waltons* and *Little House on the Prairie* are notable exceptions.

The portrayal of women is equally stereotypic and unrealistic. Long and Simon noted that of 34 female characters observed in 22 programs meant for children and families, most played comic roles or supportive wives and mothers.[39] None of the married women worked outside their homes, and, of the single women and widows who did, only two occupied positions of prestige. Those two appeared more subservient and dependent and less rational than their male colleagues. Women never occupied positions of authority, whether at home or on the job. They were usually portrayed as silly, overemotional, and dependent on husbands and boyfriends. Most were under 40 years old, well groomed, and attractive. They were concerned with their appearance, their families, and their homes. The authors concluded:

> ...this general image of women...is one of traditional sexism. The young people to whom these shows are largely or primarily aimed are not likely to gain any insights into the new roles and perceptions that many women have of themselves or want for their daughters.[40]

Minorities. Historically, stereotyping has been a prominent method in the dramatic presentation of minority characters. As Spurlock noted in 1978, this historical treatment continues.[41] The early portrayals by film and television of minority group members have moderated but not changed. We still see Asians as karate experts, as villains operating with cunning, slyness, and inscrutability, and as Charlie Chan types, supergood, superwise, self-effacing detectives. Black characters are still song-and-dance entertainers or buffoons in comedy shows, and Native Americans are still seen as alcoholics or fierce savages. The images of minorities are still negative. The Black, for instance, is victimized and limited (*Of Black America*), immature and irresponsible (*Hey, Baby, I'm Back*), foolish and stubborn (*The Jeffersons*), or mother-dominated and fatherless (*Julia, Good Times*). Although more positive depictions of minority characters have appeared in these series, in miniseries, in docudramas, in other series, and in public television programming, the standard televised drama of prime time maintains stereotyped images of minority figures. The stereotyped minorities in television drama are certainly not helpful to minority children who are striving to evolve positive self-images. Such children, already laboring under significant social and educational handicaps, are also exposed to more hours of television than white children. Nor are these images helping nonminority children and adults to accept and understand their fellow human beings.

Feelings and traits. Stereotyping television characters extends to the manner in which their feelings and personal traits are presented. Often, presumably to reinforce the story line in an unambiguous way, characters seem to possess but a single trait, such as evil power, brilliance, aggressiveness, stupidity, sexual attractiveness, weakness, and so forth. Other personal characteristics that might be obvious in any real-life person are eliminated; television's heroic characters rarely show ambivalence, confusion, a capacity to make mistakes, and so forth.

This monochromatic characterization creates false models of people, models which may be quite attractive to some viewers. In susceptible individuals this attraction can lead to a degree of identification which may be antithetical to healthy, balanced growth in the person. Monochromatic characterization, particularly when positive and negative traits are split between two separate characters, may reinforce a splitting tendency within the viewer. When the characters are split according to positive and negative traits, the viewer is encouraged to identify with the positive side and to displace or project the negative side onto others. The viewer can feel fully comfortable in doing this after having watched the heroes on a favorite program do the same thing over and over again.

Attractive characters which express only positively valued traits may also convey the implicit message that anyone who shows ambivalence, inner conflict, or confusion is less than adequate. Such a message is quite misleading about the real nature of human feelings and thoughts. The susceptible viewer may feel guilty about his or her perceived inadequacies and attempt to suppress or displace those demonstrably undesirable traits or patterns onto others in the environment.

Collins demonstrated that the presentation of a judicious mixture of human qualities in dramatic characterizations can affect the viewer.[42] He found that elementary school-age children became more aggressive immediately after seeing an aggressive televised act performed by a villain with redeeming features than they became after viewing a villain with uniformly bad motives. This might suggest that those intent on employing violence in television presentations should, after all, use monochromatic characterizations, thus discouraging children from imitation. On the other hand, encouraging viewers' identification with characters whose behavior is admirable might also be enhanced if those characters were portrayed as having a mixture of qualities. Television could thereby provide more wholesome and more accurate role models and could create more interesting programs as well.

Stripping a character of the variety of traits which make it human produces a crude stereotype so meager and unattractive as to be bad entertainment. An actor commented to one of the authors that he was bewildered and disheartened when he was offered 11 guest roles of the same type on different dramatic programs. In each case he was asked to play one-dimensional, nearly motivationless murderers. "Aside from my concern about the social consequences of such portrayals, I was baffled by the technical problem of *how* to play such characters so that anyone with discernment would want to see the program." To confine him to playing individuals who kill without evident motive was, he felt, to reduce him from an artist to a mannequin.

Dehumanization. A disturbing consequence of stereotyping is its potential for reinforcing dehumanization, a malignant and socially destructive pattern of dealing with other people. Dehumanization is a particular type of psychological defense mechanism which operates to defend the individual against intolerable feelings or unacceptable ideas, such as hostile, murderous thoughts, as well as guilt feelings about either hurting or demeaning others, or using others as displacement targets for intense feelings of frustration and anger.[43] Faced with such feelings or wishes to act, and nominally blocked by the prospect of feeling guilty or inhibited, an individual may unconsciously alter his or her perception of particular individuals or groups. By devaluing and reducing their significance as human beings, and by caricaturing them as stereotypes, the perpetrator is able to regard such dehumanized individuals as undeserving of kindness and concern. He can then treat them as if they were unfeeling, nonperson objects, suitable recipients for feelings and actions which would be intolerable and unacceptable toward fully human people.

The word *dehumanization* applies to the perceptual shift which occurs in the perpetrator, to forces or influences in the environment which provoke, promote, or reinforce such perceptions and behavior, and to the effects on the target person or group. The defense of dehumanization leaves its mark on the self as well as on the object, since the process is a circular one. Its use leads to ever-increasing degrees of dehumanization of both the self and the object.

As television drama reduces its characters to stereotypes and casts them into depreciated roles, it contributes to this process of dehumanization. Television legitimizes these dehumanized presentations by the very act of portraying them. Emphasizing single, often negative, qualities, the stereotyping process allows the viewer to countenance progressively greater insult, violence, and destruction of a character. Because the character is less than human, the viewer can drop his or her usual controlling senses of conscience and guilt. In fact, some individuals may be attracted especially to those television programs that distort, dehumanize, and destroy broad groups of people in the name of entertainment because of the psychological comfort which they gain from the character stereotyping. Presentations which dehumanize their objects of attack also tend to legitimize the violent attack itself by splitting "good" and "evil" and creating an accepting atmosphere for the violence. The fact that such presentations occur repeatedly becomes a further legitimization of the viewer's growing dehumanized views of others.

Over time, this process can lead to important changes in an individual. Diminished concern for specific individuals or groups contributes to a progressive insensitivity to a variety of feelings, especially negative ones. Such an insensitivity effectively isolates the individual by depriving him or her of a means for understanding the emotional lives and experiences of others, a capacity we call empathy. When the person has few or no empathic connections to other people, the spiraling dehumanization process is free to go unchecked. Ultimately, the process eliminates the possibility that feelings of concern for others or identification with them will interrupt or terminate destructive acts.

Dehumanization reinforces the universal tendency to project blame and fault. By silencing the voice of conscience, dehumanization prepares the way for a guilt-free attack on the external "cause" of one's discomfort and makes difficult or even impossible any acceptance of personal responsibility for one's feelings or even for one's actions. Television's use of violence to resolve conflict, of stereotyping (particularly of the victims of violence), and of monochromatic characterization, and the way these strategies promote the dehumanizing process, especially in passive and susceptible individuals, are important matters for further study.

Beyond our interest in how television encourages the dehumanization of individuals, we are concerned with its potential to lead large groups of people to engage in dehumanization of others. The Nazi propaganda films of the 1930's illustrated the power of filmed drama as a mass persuader in the hands of unscrupulous authority. Equally unscrupulous use of televised drama now could serve socially malignant purposes with vast consequences. The complicated process of stereotyping and dehumanization by groups involves contagion, displacement of responsibility to the leader or major actor, acceptance of submissiveness, and a marked decrease in discriminating judgment. Illuminating the insidious capacity of current television drama to promote the individual dehumanizing process would help us to prevent both individual and social harm. We might also, if we understood the stereotyping and dehumanization processes more completely, contribute to a better understanding of prejudice and discrimination, thereby improving the quality of contemporary social life.

REFERENCES

1. May, Rollo. "Love and Will." *Psychology Today*, 1969, *3* (3), p. 17-61.
2. Smythe, D.W. "Reality as Presented by Television." *Public Opinion Quarterly*, 1954, *18*, 143-156.
3. Comstock, G., Chaffee, S., Katzman, N., McCombs, M., and Roberts, D. TELEVISION AND HUMAN BEHAVIOR. New York: Columbia University Press 1978, p. 64.
4. Gerbner, G., Gross, L., Eleey, J.F., Jackson-Beeck, M., Jeffries-Fox, S., and Signorielli, N. VIOLENCE PROFILE #8: "Trends in Network Television Drama and Viewer Conceptions of Social Reality," *Journal of Communication*, 1967-76. Philadelphia: Annenberg School of Communications, University of Pennsylvania, 1977.
5. Gerbner, G., Gross, L., Morgan, M., and Signorielli, N. "The 'Mainstreaming' of America: Violence Profile No. 11." *Journal of Communication*, 1980, *30* (3), 10-29.
6. Gerbner, G., Gross, L., Morgan, M., and Signorielli, N. "The 'Mainstreaming' of America: Violence Profile No. 11, *Journal of Communication*, 1980, *30* (3), 11.

7. Comstock, G., Chaffee, S., Katzman, N., McCombs, M., and Roberts, D. TELEVISION AND HUMAN BEHAVIOR. New York: Columbia University Press, 1978, pp. 64-79.
8. Gerbner, G., Gross, L., Eleey, M.F., Jackson-Beeck, M., Jeffries-Fox, S., and Signorielli, N. VIOLENCE PROFILE #8 "Trends in Network Television Drama and Viewer Conceptions of Social Reality," *Journal of Communication*, 1967-76. Philadelphia: Annenberg School of Communications, University of Pennsylvania, 1977, 1.
9. Somers, Anne R. "Violence, Television and the Health of American Youth." *New England Journal of Medicine*, 1976, *294* (15), 813.
10. Baker, R.K., and Ball, S.J. "The Television World of Violence." In R.K. Baker and S.J. Ball (Eds.), VIOLENCE AND THE MEDIA: A STAFF REPORT TO THE NATIONAL COMMISSION ON CAUSES AND PREVENTION OF VIOLENCE. Washington, D.C.: Government Printing Office, 1969, pp. 311-339.
11. Bandura, A. "Influence of Models' Reinforcement Contingencies on the Acquisition of Imitative Responses." *Journal of Personality and Social Psychology*, 1965, *1*, 589-595; Bandura, A. AGGRESSION: A SOCIAL LEARNING ANALYSIS. Englewood Cliffs, N.J.: Prentice-Hall, 1973.
12. Liebert, R.M., Neale, J.M., and Davidson, E.S. THE EARLY WINDOW: EFFECTS OF TELEVISION ON CHILDREN AND YOUTH. New York: Pergamon Press, 1973.
13. "What TV Does to Kids." *Newsweek*, 21 February 1977, p. 68.
14. Wertham, Frederic, quoted in Liebert, R.M., Neale, J.M., and Davidson, E.S. THE EARLY WINDOW: EFFECTS OF TELEVISION ON CHILDREN AND YOUTH. New York: Pergamon Press, 1973, p. 34.
15. Redl, Fritz. Personal Communication, 1978-1979.
16. Dorr, Aimee. "The Effects of Television Violence on Children's Behavior." Paper presented at annual meeting of the American Orthopsychiatric Association, April 11-16, 1977.
17. Liebert, R.M. "Television and Social Learning: Some Relationship Between Viewing Violence and Behaving Aggressively." In J.P. Murray, E.A. Rubinstein and G.A. Comstock (Eds.), TELEVISION AND SOCIAL BEHAVIOR. VOL 2. TELEVISION AND SOCIAL LEARNING. Washington, D.C.: Government Printing Office, 1972, pp. 181-201. The five studies are described in Volume 3 of TELEVISION AND SOCIAL BEHAVIOR: TELEVISION AND ADOLESCENT AGGRES-

SIVENESS, (G.A. Comstock and E.A. Rubenstein, Eds.) include: McIntyre, J.J., and Teevan, J.J., Jr., "Television Violence and Deviant Behavior," pp. 383-435; McLeod, J.M., Atkin, C.K., and Chaffee, S.H. "Adolescents, Parents and Television Use: Self-Report and Other-Report Measures from the Wisconsin Sample," pp. 239-313; Robinson, J.P., and Bachman, J.G., "Television Viewing Habits and Aggression," pp. 372-382; Dominick, J.R., and Greenberg, B.S. "Attitudes Toward Violence: The Interaction of Television Exposure, Family Attitudes and Social Class," pp. 314-335; and Lefkowitz, M.M., Eron, L.D., Walder, L.O., and Huesmann, L.R., "Television Violence and Child Aggression: A Follow-Up Study," pp. 35-135.

18. Lefkowitz, M.M., Eron, L.K., Walder, L.O., and Huesmann, L.R. GROWING UP TO BE VIOLENT: A LONGITUDINAL STUDY OF THE DEVELOPMENT OF AGGRESSION. New York: Pergamon Press, 1977, pp. 115-116.

19. Belson, William. TELEVISION VIOLENCE AND THE ADOLESCENT BOY. London: Teakfield, 1978.

20. Liebert, R.M., Cohen, L.A., Joyce, C., Murrel, S., Nisonoff, L., and Sonnenschein, S. "Predispositions Revisited (Symposium on Effects of Television). Review of Heller & Polsky: 'Studies in Violence and Television,' New York: ABC, 1975." *Journal of Communication,* 1977, 27 (3), pp. 217-221.

21. Comstock, G. *TV Portrayals and Aggressive Behavior.* Santa Monica, Calif.: Rand Corporation, 1976, P-5762; Liebert, R.M., Neale, J.M., and Davidson, E.S. THE EARLY WINDOW: EFFECTS OF TELEVISION ON CHILDREN AND YOUTH. New York: Pergamon Press, 1973; Liebert, R.M., and Schwartzberg, N.S. "Effects of Mass Media." *Annual Review of Psychology,* 1977, 2, 141-173; Rothenberg, Michael. "Effects of Television Violence on Children and Youth," *Journal of the American Medical Association,* 1975, 234 (10), 1043-1046, Dec. 8.

22. Belson, William. TELEVISION VIOLENCE AND THE ADOLESCENT BOY. London: Teakfield, 1978.

23. Comstock, G. *TV Portrayals and Aggressive Behavior.* Santa Monica, Calif.: Rand Corporation, 1976, P-5762; Drabman, R.S., and Thomas, M.H. "Does TV Violence Breed Indifference?" *Journal of Communication,* 1975, 25 (4), 86-89; Drabman, R.S., and Thomas,

M.H., "Does Watching Violence on Television Cause Apathy?" *Pediatrics,* 1976, *57,* 329-331; Cline, V.B., Croft, R.G., and Courrier, R.G. "Desensitization of Children to Television Violence." *Journal of Personality and Social Psychology,* 1973, *27,* 360-365.
24. Gerbner, G., and Gross, L. "Living with Television: The Violence Profile." *Journal of Communication,* 1976, *26* (2), 173-199; Gross, L. "The Scary World of TV's Heavy Viewer." *Psychology Today,* 1976, 9 (12), 41-45, 89.
25. Berkowitz, L., and Rawlings, E. "Effects of Film Violence on Inhibitions Against Subsequent Aggression." *Journal of Abnormal and Social Psychology,* 1963, *66,* 405-412.
26. Gerbner, G., and Gross, L. "Living with Television: The Violence Profile." *Journal of Communication,* 1976, *26* (2), 173-199; Gross, L. "The Scary World of TV's Heavy Viewer." *Psychology Today,* 1976, 9 (12), 41-45, 89.
27. Gerbner, G., Gross, L., Morgan, M., and Signiorelli, N. "The 'Mainstreaming' of America: Violence Profile No. 11." *Journal of Communication,* 1980, *30* (3), 10-29.
28. Doob, A.N., and McDonald, G.E. "Television Viewing and Fear of Victimization: Is the Relationship Causal?" *Journal of Personality and Social Psychology,* 1979, *37* (2), 170-179.
29. "Violence on TV: Why People Are Upset." *U.S. News & World Report,* 29 October 1973, pp. 33, 36.
30. Diener, E., and DeFour, D. "Does Television Violence Enhance Program Popularity?" *Journal of Personality and Social Psychology,* 1978, *36* (3), 333-341.
31. Gorney, Roderic, Loye, David, and Steele, Gary, "Impact of Dramatized Television Entertainment on Adult Males." *American Journal of Psychiatry,* 1977, *134* (2), 170-174.
32. Rothenberg, Michael. "Effects of Television Violence on Children and Youth." *Journal of the American Medical Association,* 1975, *234* (10), 1043-1046, Dec. 8.
33. DeFleur, M.L. "Occupational Roles as Portrayed on Television." *Public Opinion Quarterly,* 1964, *28,* 64.
34. Seggar, J.F., and Wheeler, P. "World of Work on TV: Ethnic and Sex Representation in TV Drama." *Journal of Broadcasting,* 1973, *17,* 201-214.
35. Head, S.W. "Content Analysis of Television Drama Programs."

Quarterly of Film, Radio and Television, 1954, *9,* 175-194.
36. DeFleur, M.L. "Occupational Roles as Portrayed on Television." *Public Opinion Quarterly,* 1964, *28,* 57-74.
37. Seggar, J.F., and Wheeler, P. "World of Work on TV: Ethnic and Sex Representation in TV Drama." *Journal of Broadcasting,* 1973, *17,* 201-214.
38. Tedesco, N.S. "Patterns of Prime Time." *Journal of Communication,* 1974, *24,* 119-124.
39. Long, M.L., and Simon, R.J. "The Roles and Statuses of Women on Children and Family TV Programs." *Journalism Quarterly,* 1974, *51,* 107-110.
40. Long, M.L., and Simon, R.J. "The Roles and Statuses of Women on Children and Family TV Programs." *Journalism Quarterly,* 1974, *51,* 110.
41. Spurlock, Jeanne. "Television, Ethnic Minorities and Mental Health: An Overview." Paper read at Conference on the Socialization of the Minority Child, Los Angeles, University of California, April 27-28, 1978.
42. Collins, W.A. ASPECTS OF TELEVISION CONTENT AND CHILDREN'S SOCIAL BEHAVIOR. Minneapolis, Minn.: University of Minnesota, Institute of Child Development, 1974; Collins, W.A., "The Developing Child as Viewer." *Journal of Communication,* 1975, *25* (4), 35-44.
43. Bernard, V.W., Ottenberg, P., and Redl, F. "Dehumanization: A Composite Psychological Defense in relation to Modern War", *in Behavioral Science and Human Survival,* M. Schwebel (ed.). Palo Alto, CA: Science and Behavior Books, 1965.

5
THE PROGRAMS: CHILDREN'S TELEVISION

Cartoons and the new educational television programs are of special interest to us because they are consciously and carefully designed to reach the young child. Because of this, they will theoretically have a considerable impact on their young viewers. Both types of programs fall within our definition of drama, since they commonly depict stories played by actors, human or animal. The dramatic device, as we have noted, is of special concern because of its capacity to engage viewer involvement and, thereby serve as a stimulus to behavior. But because the cartoons and educational programs are experienced as less "real" than the live dramas, they have the possibility of avoiding some of the problems the live dramas pose for the child viewer. On the other hand, they have some problems of their own. We touch on both their prospects and their problems.

CARTOONS

Cartoons are a substantial part of the television children watch, espcially on the Saturday morning "kid-vid ghetto." They are also a sizable quantity of what limited special programming there is for children. Because of this and because they are so popular all over the world, especially with preschool children, cartoons warrant serious consideration.

Children's interest in cartoons stems in part from their appropriatcness to developmental needs at various stages. Between the ages of one and three, the child develops a working language. As he or

"We don't like the space walk because they have it on Saturdays and we don't get to see cartoons."

Photo, courtesy of Bill Owens

she does so, ideas of "person" begin to emerge through knowledge of mother, father, and self. During this early period, the toddler moves from the simple patterned play of light and sound to experiences which have increasing meaning and which are sources of symbolic and conceptual learning. In their audiovisual format, cartoons retain aspects of the light and sound play, while their drama and dialogue reinforce the developing language and sense of persons. As a medium, the cartoon relies heavily on visual symbolism, potentially a useful support to the viewer's developing conceptual ability.

Cartoons are perceptually attractive, easily catching the attention of both the ear and the eye. Even more important is their appeal to the age-old desire to hear stories that inform one about both the world and the inner life of the story teller. The line drawing in cartoons eliminates irrelevant and distracting detail, and the simple images are easy for a young child to assimilate. As image and event are easier to distinguish from background than they are in ordinary live film, the young child can more easily grasp the point of a story.

Of special significance is the interesting capacity of cartoons to display frightening, threatening, or scary events without evoking fear or anxiety in the viewer. In a group of 50 children, 73% of those four to eight and 83% of those nine to 12 described violent behavior in animated cartoons as *not* violent.[1] In contrast, 68% and 65% of the two age groups, respectively, described violent behavior in adult westerns as violent. All of the children judged films of the Vietnam War to be violent. Those who did not describe the violent behavior as violent in either format explained that the programs were "funny" or "make-believe," as if they only considered violence to be violent when it was in a "real" context.

An interesting natural experiment connected with *Star Trek,* the television science fiction series, is also illustrative of this point. During 1973-1974 a dramatized portrayal by actors and a cartoon version of *Star Trek* were both available on weekend programming, sometimes with identical plots. When asked which he preferred, an avid nine-year-old fan responded, "If it's a real scary program, I'd rather watch the cartoon version; it's not so real."

Human dramatization is experienced as more real, richer in information and feeling, and capable of generating more empathy and identification. It can, therefore, be less appealing or more threatening to a young child who is unable to handle its complexity.* Since most children see cartoons as make believe, different expectations and different rules apply. The child can experience the unthinkable at a distance and deal with it more easily. A sudden unpredictable cataclysm, an engulfing monster, a murderous witch or villain, a violent denouement in which the young protagonist plays a part, or a rivalrous physical struggle—the child's interest in watching these subjects comes from his own wish to achieve mastery over one or another developmental problem. Using the cartoon fantasy can help.

Cartoons depict an "as if" world of pseudoreality and indulgence. Children can have multiple identifications with all of the characters in the drama. They can, thereby, accept the threat of extinction in one character and survival in another, or laugh at the stupidity in a character, rather than fear it in themselves. Switching identifications, the child can relieve conflict and choose a different threshold of excitement. The viewer can be both victim and victor, lover and loved, outcast and group member, or alive and dead, according to his or her feelings and his or her ability to tolerate their expression.

Cartoons permit children to select from explicitly nonreal presentations of emotional issues those which they are able to tolerate. Yet sometimes, if the cartoon content too closely parallels an issue about which a child has considerable unconscious fear or guilt, the protective aspects of the line drawing and the reduced sense of reality may not be enough to neutralize the strong feelings. Vulnerable seven- or eight-year-olds may feel intense anxiety when their own developing senses of reality are still relatively unstable and their senses of good and bad are more absolute than what they are viewing.

---

* See clinical example on page 63

But the risks are not only in the emotional responses a child may have in the short run. While the child is amused by the cartoon, assisted in fantasy development, or protected from frightening things, he or she is also learning from the cartoon. Cartoons, intentionally or unintentionally, teach about people, feelings, and the outside world. Depicted in abbreviated form and distorted by the otherwise beneficial simplification of cartooning, characters are grossly stereotyped. Animals take on human personality traits, manifesting single, unmitigated feelings and traits—love, hate, competition, cleverness, strength, or cowardice. The oversimplified stereotype appears invariably in a given cartoon character and often contrasts with its polar opposite which is manifest in a rival. What the child learns about people through these characters is distorted and unreal. If cartoons are a major source of information, this distribution is likely to feed into stereotyping and dehumanization at later ages.

Our concern about violence extends to the cartoons. So much of what happens between cartoon figures involves violence that cartoons sometimes seem to be nothing but a connected string of violent events. For example, on a particular Saturday morning:

> ...a mean, gray coyote barrels down the highway in hot pursuit of the roadrunner. The coyote slams into a tree. Seconds later, he falls off the edge of a cliff and gets flattened by a two-ton boulder. A slab of road pavement flips over and buries him. A piano wired with dynamite blows up in the coyote's face. He goes flying through the air and lands with a mouthful of keys. The entire sequence lasts less than three minutes.[2]

Such a sequence may undergo a dozen repetitions in a given cartoon and also recur in succeeding cartoons. Child viewers may watch the same programs ten times in a year. Yet these repeatedly battered animals (people) never seem to suffer. Like the Bobo-the-Clown doll, they keep coming back for more, undeterred, uninjured, unaffected. As in the dramas with live actors, the real consequences of violence are not shown.

Cartoons do entertain preschoolers. Given the opportunity to watch them, most young children will. In cartoon form, the characters and the presentations do have a reduced emotional impact on the young child. Although some cartoons may be too frightening or too stimulating, the effect of the moment is probably not as significant as the overall amount of time spent viewing. As we have noted previously, the balance between time spent watching television and other learning experiences, especially those with other people, is especially important for this age group.

When should parents be concerned about the cartoons their children watch? When some or all of these circumstances prevail:

- when television cartoons are generally preferred to other available activities (playing with other children, being read to, playing alone);
- if the preschool child who is routinely allowed to watch also shows little interest in other activities;
- if environmental options for stimulation and gratification are so limited that the child has little choice;
- if either or both parents use television in place of involvement with the child rather than as an opportunity for shared experience;
- if older siblings force upon younger siblings program choices they cannot handle;
- if the child prefers cartoons past the age when other program choices are more age appropriate;
- when the child's play mirrors the cartoon's content excessively and repeatedly;
- when the child has nightmares or other regressive symptoms centered on the cartoon content;
- when the child's behavior changes to imitate the mannerisms, speech, and style of the cartoon characters for prolonged periods.

Photo, courtesy of Shearer Visuals

EDUCATIONAL PROGRAMS

How can television educate?

Building knowledge and skills. Television is undoubtedly a teacher. As Schramm has observed, because television is "perhaps richer and denser than any other instructional medium, children have almost invariably shown learning gains, often large ones."[3] Using scripts for television plays as texts, some Philadelphia schools have demonstrated dramatic improvement in the reading skills of the children involved. Much of the success here was due to the design of the program as a coordinated activity for the whole family and to the active involvement of the parents.

In other parts of the country, the recognition that regularly watched, prime time television offers opportunities for learning has led to a number of notable initiatives. Prime Time Television, a Chicago-based nonprofit organization, has developed a series of study guides to illustrate the general principles of prime time teaching. Use the screen and the dramatic show, they recommend, to get the students' attention. Then engage their intelligence with questions, study guides, and scripts to be read as homework. These and other teaching approaches have emphasized logic and language skills, literary and dramatic elements, conflict development and resolution, and civil rights and constitutional law, to identify a few important topics.

Promoting healthy psychological development. Clearly, with appropriate programming and instructional guidance, television can teach. But can it educate? Can it draw a person out and help a person to grow psychologically? We have already discussed its possible harmful effects on psychological well-being.

The possibilities of television for aiding the psychological growth and development of children are less apparent to many than the hurt we have observed in some individuals and the deleterious aspects noted in the literature. Some exceptional efforts are, however, worth noting.

A series of 15-minute episodes called *Inside/Out,* available to public schools as instructional television, portrays youthful actors in a variety of actual and potential conflicts, dealing with secretiveness, rejection, vengefulness, or jealousy. Under the carefully nonjudgmental leadership of a teacher, classroom viewers discuss the alternative outcomes, the possible consequences, and the solutions they might have considered had they been the protagonists. The children learned more from the discussion period than from the television program, and the discussion itself was especially focused and engaging because it followed the televised scenarios. Having an adult mediator also seemed an important contribution to the psychological benefit the children received.

Singer and Singer confirmed the vital role of the adult mediator in a different setting with a different age group.[4] They studied the impact of television on the imaginative play of younger children. In the study, four-year-old children viewed a *Mister Rogers' Neighborhood* segment with and without a mediator. As evidence of imagination, researchers observed whether or not a child used an object or toy in a way different from what its appearance would ordinarily suggest. They found that viewing the program alone was less stimulating to imaginative play than might have been hoped for. The four-year-old children could not sit still throughout the half-hour program. When the adult was present during the television viewing period, concentration improved, interaction increased, involvement with the program was greater, and evidence of imaginative play was greater. The researchers inferred that children of this age are most susceptible to the influence of an adult who can engage them directly and reinforce their own responses immediately, something which television alone, no matter how positive, cannot do. The implications for parental or other adult intervention are clear. Apparently, the adult can and should play a powerful mediating role, perhaps offsetting the harmful viewing material and helping the child to enjoy more of the potential benefits from apparently positive programming. Thus, responsible adult involvement with children's viewing is absolutely necessary.

Fostering positive social relations. Whether television now teaches or could teach socially beneficial behavior has been and continues to be a focus of considerable attention. Examining more than 300 adult- and child-oriented television programs in 1974, Davidson and Neale recorded many examples of behavior they defined as generally socially valued, especially in the categories of altruism and sympathy.[5] Resistance to temptation and control of aggressive impulses were, however, virtually absent. A more detailed look at typical Saturday morning children's television (entirely cartoons) also demonstrated some socially valued behavior.[6] Of interest, though, was their finding that the incidence of socially positive acts (altruistic sharing, helping, and cooperation) was the same as that of aggressive acts—about six per half-hour program.

Friedrich and Stein were among the first to study the socially beneficial possibilities of the television medium.[7] Examining a group of 93 preschool children, they demonstrated that children who watched *Mister Rogers' Neighborhood* showed measurable increases in positive interpersonal behavior as compared with those children who were exposed to aggressive cartoons. Those who benefited also showed higher levels of task persistence and somewhat higher levels of rule obedience and delay tolerance. They did observe, however, that the substantial benefits were limited to children from lower socioeconomic backgrounds; children from higher status families did not realize comparable gains. Those who benefited also tended to be children with above-average intelligence.

Using episodes from the *Lassie* series, one team of investigators illustrated that models of helpfulness can effectively stimulate similar behavior among the viewers.[8] They also demonstrated that a television spot announcement specifically designed to illustrate sharing behavior significantly increased cooperation among the children who watched and that this increase occurred after as little as a single exposure.[9]

Stein and Friedrich observed Headstart children watching a children's television program both by itself and accompanied by special materials, attention, and training.[10] Children who merely viewed the program showed only slight effects. Those children given supplemental attention and relevant play materials by the teachers showed gains in both positive interpersonal behavior and in psychosocial skills. They were more imaginative at play and used more role-playing fantasy. This reinforces the need for adult involvement at the time of viewing, whether by teachers or parents.

The importance of the adult as a model was underscored in the positive results from a study by Collins.[11] Among fourth, seventh, and tenth graders, Collins found that exposure to televised models behaving in positive ways produced greater helpfulness and support for peers than exposure to aggressive models or to programming in which no relevant models were present.

In some programming for children, producers have consciously developed positive identification models. *Zoom* is of particular interest in this regard. A group of preadolescent children are part of the program. Younger viewers generally take interest in the activities of age groups a few years older than themselves. In *Zoom* the televised older children are involved in activities and challenges the younger child is just beginning to know and understand. This scenario excites the younger children. Precocious preschoolers and children in the lower grades of grammar school are especially attracted by the activities of the *Zoom* kids. Their attentive response points up again the importance of appropriate models, whether they be live adults, televised adults, or child actors playing appropriate roles. Youthful roles and models with significant human dimensions are few in today's daytime reruns of the many adult situation comedies, even those with central child actors.

Beyond these particular examples of socially beneficial programming and related research, we believe that some programs, by virtue of their choice and portrayal of characters, have potential benefits. Television programs which present a diversity of persons in successful roles and activities, which are not stereotyped, tend to be beneficial. Programs for bilingual children, programs showing black, Oriental, or handicapped children in realistic, nonstereotyped roles, and even programs showing children with single parents, can be helpful. Such minority children in the viewing audience may feel less singled out when they can watch the successful adventures of children like themselves.

As well as the direct benefits that children can gain from positive programs created especially for them, children can benefit from the positive effects that some programs for adults foster in the home. The previously mentioned experimental field study of 183 married couples demonstrated that socially positive programs produced significantly less aggressive moods and hurtful behavior among the adults than did the programs with a high violence content. Such positive effects enhance the emotional climate of the home, on which children mainly depend for support as they move through their normal but often troublesome developmental sequences. Conducted in the natural setting of the home, this study yielded results parallel to the findings of the many laboratory studies that have confirmed the beneficial effects of socially positive programs. Moreover, the fact that the study was in the natural setting counters the legitimate contention that the findings of laboratory studies cannot be generalized to the real-life condition.

Children's educational programs

Soon after the beginning of commercial television in this country, programs especially intended for children appeared, though tentatively and in small numbers. *Captain Kangaroo* (Bob Keshan) led the way, and in 1954 Fred Rogers began a short program on Station QED in Pittsburgh. Entitled *Children's Corner* the program subsequently expanded to national television as *Mister Rogers' Neighborhood*. These initial efforts were given an enormous boost with the appearance of *Sesame Street* in November 1969, and in its wake have come other programs of significance, offering other perspectives and methods—*Electric Company, Mi Casa, Su Casa, Big Blue Marble,*

Villa Alegre, Carra Scolendas, and many others. Worth noting is the fact that some of these programs have benefited from extensive advice from mental health professionals. Psychologists Gerald Lesser, Ed Palmer, and others advised on *Sesame Street,* and *Mister Rogers* has had the advice of psychologist Margaret McFarlane.

Sesame Street was an experimental program, developed by the Children's Television Workshop with the aid of grants from the Carnegie and Ford Foundations. Since it was developed without the usual need of children's television to sell products, great innovation and flexibility resulted, compared with the usual cartoon fare. A pioneer in the development of new techniques and methods intended to educate young viewers, *Sesame Street* demonstrated that positive learning on several levels—cognitive, affective, and attitudinal—can result from a format that is both entertaining and engaging.[12] Partly because this program was innovative in a number of ways, partly because it began at a time when educational television was held in low esteem and new initiatives were badly needed, and partly because the program assertively undertook to teach material felt to be important to young children, *Sesame Street* was widely hailed as a milestone in education as well as in television. The program offered a means of preparing preschoolers for formal education. For children whose disadvantagement threatened lifelong learning handicaps, *Sesame Street* promised to intervene in the cycle.

Sesame Street did break ground in many areas. The program developed methods for enhancing perceptual and symbol-recognition skills. It broadened vocabulary and introduced many concepts new to preschool children. The program attacked racism through its careful casting and role playing. It engaged viewer participation through adroit use of studio audiences similar in age to the television audiences, and it presented new material in familiar contexts, easing the transition from the known to the unfamiliar.

Although *Sesame Street* is in many ways an oasis in the desert of television, its actual impact on the educational growth of young children—especially disadvantaged children—is in doubt, according to the published evidence. An early study by Ball and Bogatz suggested that children encouraged to view *Sesame Street* did watch and did demonstrate learning gains over relatively short (six-month) periods.[13] Cook and others challenged those findings in a subsequent analysis.[14] In effect, while the latter study team found that children who viewed more did gain more, they also found that much of the gain was caused not by high viewing itself, but by natural maturation and environmental factors. They found, moreover, that the more socially advantaged children made larger gains. Thus, rather than closing the gap between the disadvantaged children and their more fortunate peers, *Sesame Street* seemed unintentionally to be increasing it. In fact, Liebert observes, one study showed that the first year of *Sesame Street* produced gains only for advantaged childen.[15]

At the same time, in spite of its failure to live up to the high expectations that surrounded its inauguration, *Sesame Street* has generated new interest in the educational potential of television, and, as noted, has inspired the development of a number of other significant programs. These, like *Sesame Street,* have much promise, which time and experience will assess.

Some concerns. We are, however, concerned about the approaches used in some of these programs. The rapid-fire techniques of presentation used do seem to increase the retention and recall of sounds and words among children who might not have learned them otherwise, but it is not clear that long-range learning goals are best served when the child learns the sounds and words in contexts unrelated to readiness for reading. Too many concepts, perhaps, are introduced too quickly. To facilitate reinforcement and useful integration with other learning, concepts should be repeated in different contexts. Rapidly paced, marginally comprehensible sequences can be confusing and may be damaging as well. These sequences are intended to attract and to hold attention, but they show no concern for the slower paced processes of reflection and integration and little recognition of the varying needs and abilities of different individuals in various developmental phases.

Make a Wish, for instance, frequently alternates a several-minute segment of informative narrative, such as a person narrating slowly, with several minutes of rapid-fire, multisensory programming. In the *verbatim* transcript of the auditory part of a two-minute segment on mountains, a double dash (--) stands for the word "mountain":

> Mountain, -- goat, -- laurel, -- summit, -- bear, rugged, Russia --, range, Tetons, Rushmore; Black Hills, history, heads of our presidents carved out granite cliff, chip off the old block; Mountain! (A poetic description of mountains follows for a few seconds at a slower pace. Original pace returns.)
> Mountain standard time, -- passes, -- peak, -- peek, -- states, Montana, -- lion, roar, reaches for stars and make men reach.
> (There then follows a slow-paced sequence for several minutes of a man talking of his feeling for the mountains around Taos, New Mexico.)

The rapid stimulation is not only in the script. It is multimodal. All at the same time, visual image, sound, and print stimulate young viewers to grasp complexity as quickly as they can. Very likely, this barrage confuses the child and obscures the message. An example often cited comes from commercials aimed at children. An exciting picture of an attractive toy appears, and an enthusiastic announcer describes it. Meanwhile, the words "batteries not included" or "each part sold separately" are shown in fine print, there to be seen but neither perceived nor understood by the young child.

On the basis of analysis of the formal features of children's television programming, Huston and others concluded that "[commercial children's programs] would appear to be intended for a particular type of child viewer whose attention must presumably be captured and held by constant action, change, noise, visual onslaught, as well as slapstick violence."[16] Concerning the effects of this rapid, multimodel approach, Singer reports that the high-frequency television viewers among their preschool subjects use more nouns and adjectives in their spontaneous speech during play, while low-frequency viewers use more verbs and adverbs.[17] High-frequency viewing, he suggests, is conducive to simple and discontinuous patterns of speech.

As in so much other television, most children's programs rarely present activities in which the audience can participate. Two notable exceptions are oriented to the youngest age groups and were carefully based on knowledge of child development. *Watch Your Child* and *Mister Rogers' Neighborhood* both include experiences to foster imitation and participation by the child at home and are organized to promote the child's assimilation of cognitive and affective stimuli. Mr. Rogers is warm, gentle, friendly, and relaxed. He moves in a slow and reflective way through the several experiences of the half-hour program. The child viewer is encouraged to absorb, reflect upon, participate in, and finally integrate the experiences he is sharing.

Mister Rogers' Neighborhood, again developed outside the structures of commercial programming, has been shown in several studies to be an effective teacher of socially beneficial behavior.[18] Children who initially rated low on measures of imagination made significant gains over a two-week period of viewing *Mister Rogers,* as compared with children who viewed *Sesame Street*. The authors postulated that the clear-cut distinction between the reality and the make-believe kingdom on the Mister Rogers program stimulated imagination. The program also stimulated positive social behavior in those children: they tended to show quite positive feelings, had a high level of concentration, engaged in many interactions with their peers, and cooperated with other children. These data suggested that while *Sesame Street* did entertain children and hold their attention, the less verbally intelligent children were unlikely to assimilate and retain the material. *Mister Rogers,* on the other hand, seems to have special advantages in facilitating comprehension and in building up constructive, imaginative play among children who are initially less imaginative. This format appears to be a good one for developing future programs.

Educational programs can lead to relatively rapid increases in personally and socially constructive behavior among preschoolers. *Sesame Street* and its followers have pioneered in the deliberate effort to offer inspiration and love, relying on a sophisticated combination of art and science. But as with any other television program, it is important to respect the cognitive, emotional, and social needs and abilities of the (young) viewers. Moreover, if educational programs are to be helpful to those children who need extra stimulation and help—and are most likely to be heavy television viewers—the needs and abilities of these children, in particular, must guide the design of programs.

A programming need: promoting mental health

As psychiatrists, we have a special interest in the possibility that television can enable young viewers to learn more about their affective natures, or feelings. Television might help people learn about anger, what generates it, and how to manage it in oneself and in others. Perhaps it could help people to understand empathy and how to develop and use it. The problems that invariably emerge in close relationships and how they might be managed are also areas television programming could address. The many other important possibilities in this area might include learning how to deal with conflict and differences in constructive ways and how to deal with negative feelings such as hostility, envy, jealousy, loneliness, inferiority, inadequacy, and fear of rejection in oneself and others.

Dramatic presentations designed to entertain often illustrate these psychological issues and thereby teach something about them. But this teaching is secondary to the main dramatic purpose and, as we have observed, is often misleading. Television might systematically illustrate the ways the common interpersonal situations can be approached and the consequences of various approaches. A kind of trial-action-in-fantasy played out for the viewer would be useful to children (and adults) and could also be entertaining.

Despite the common view that human emotions are the source of a great deal of interpersonal conflict, very little has been done in television and elsewhere to teach people about the emotions and their management. People assume we learn enough of these things in the course of living. Or, because teaching this material to children is so difficult, it is neglected in favor of easier subjects. Consequently, curriculum is undeveloped, and adaptations for television are practically nonexistent.

An unusual and apparently effective curriculum is that which Spivack and Shure have developed and tested.[19,20] Aiming to enhance young children's ability to think through ways to solve interpersonal problems, the program includes a curriculum for teaching preschoolers and kindergarten children personal and interpersonal problem-solving skills. The curriculum helps the children to imagine or conceptualize various approaches to getting what they want (a block from another child, for example) and the consequences of each approach (hitting the child to get it, for example). Extensive tests indicate that, among Head Start children, those who most successfully acquire the problem-solving skills are superior in overall classroom adjustment. On follow-up testing in subsequent years, they appear to have retained the advantage. Findings also suggest that children who are inferior problem solvers are more likely than the superior problem solvers to give forceful, antisocial solutions, such as "grab it" or "hit him." The researchers conclude that the child's choice of an effective problem-solving approach does relate significantly to his or her ability to conceptualize alternative solutions. What needs to be tried now is to adapt this curriculum to television programming.

One approach is demonstrated in a puppet show on *Mister Rogers' Neighborhood*. Chipmunk handpuppets representing mother, brother, and sister play a short segment about a problem of having to do household chores. The handpuppet characters of brother and sister leave no doubt about how they feel about an assignment of picking up their rooms. They demonstrate rivalrous feelings for each other, which complicates getting the unwelcome task done. This skit serves as a basis for later material in which both of them see social benefits to the family of doing the chore: they find that all three family members can work together on a common task, even when they have strong negative feelings. Promoting the child viewer's identification with the handpuppet, the skit acknowledges the reality of negative feelings in the family setting, but it also shows that successful work with benefits for all is still possible.

A program with similar goals for older children is *The Big Blue Marble*. The actors, children of school age, play roles in conflicts that are particularly relevant to school-age children. Being part of a birthday celebration is the focus of one such episode. It demonstrates the feelings that surround giving and getting presents and the common feelings of envy and rivalry in relation to the birthday child. Some resolution of the tensions occurs when a kindly uncle leads a conversation among the children. The episode illustrates an experience common to all children but one which is rarely acknowledged or discussed. Showing that the feelings are legitimate and how their negative potential can be reduced or eliminated, the episode is an example of good mental health education through television.

Consistent with common sense, we believe that television presentations can exert positive as well as negative effects, and that careful dramatic and educational programming can facilitate children's personal psychological and interpersonal growth. It can also teach mental health knowledge and skills directly. But we must demonstrate that these effects do indeed occur, that programming can promote healthful processes and behavior. Imaginative programming has virtually limitless possibilities for personal and social learning. Such an exciting prospect deserves a considerable investment in media research, experimentation, and development.

REFERENCES

1. Snow, R.P. "How Children Interpret TV Violence in Play Context." *Journalism Quarterly,* 1974, *51,* 13-21.
2. Lublin, Joann S. The Television Era, *Wall Street Journal,* 19 October 1976, p. 1.
3. Schramm, W., Lyle, J., and Parker, E.B. TELEVISION IN THE LIVES OF OUR CHILDREN. Stanford, CA: Stanford University Press, 1961.
4. Singer, J.L., and Singer, D.G., "Can TV Stimulate Imaginative Play?" *Journal of Communication,* 1976, *26,* 74-80.
5. Davidson, E.S., and Neale, J.M. "Analyzing Prosocial Content on Entertainment TV." Paper presented at the American Psychological Association, New Orleans, La., September 1974.
6. Stein, A.H., and Friedrich, L.K. "Television Content and Young Children's Behavior." In G.A. Comstock and E.A. Rubinstein (Eds.), TELEVISION AND SOCIAL BEHAVIOR. Vol. 2. Washington, D.C.: Government Printing Office, 1972, pp. 202-317.
7. Friedrich, L.K., and Stein, A.H. "Aggressive and Prosocial Television Programs and the Natural Behavior of Preschool Children." *Monographs for the Society for Research in Child Development,* 1973, *38* (4), Serial #151.
8. Sprafkin, J.N., Liebert, R.M., and Poulos, R.W. "Effects of a Prosocial Televised Example on Children's Helping." *Public Opinion Quarterly,* 1975, *21,* 119-126.
9. Liebert, R.M., Sprafkin, J.N., and Poulos, R.W. "Selling Cooperation to Children." In W.S. Hale (Ed.), PROCEEDINGS OF THE 20TH ANNIVERSARY CONFERENCE OF THE ADVERTISING RESEARCH FOUNDATION, New York: Advertising Research Foundation, 1974, p. 108.
10. Stein, A., and Friedrich, L. "Impact of Television on Children and Youth." In E. Hetherington (Ed.), *Review of Child Development Research,* Vol. 5. Chicago: University of Chicago Press, 1975.
11. Collins, W.A. ASPECTS OF TELEVISION CONTENT AND CHILDREN'S SOCIAL BEHAVIOR. Minneapolis, Minn.: University of Minnesota, Institute of Child Development, 1974.
12. Lesser, Gerald S. CHILDREN AND TELEVISION: LESSONS FROM

SESAME STREET. New York: Random House, 1975.
13. Ball, S., and Bogatz, G.A. THE FIRST YEAR OF SESAME STREET. Princeton, N.J.: Educational Testing Service, 1970.
14. Cook, T.D., Appleton, H., Conner, R., Shaffer, A., Tamkin, G. and Weber, S.J. SESAME STREET REVISITED: A STUDY IN EVALUATION RESEARCH. New York: Russell Sage Foundation, 1975.
15. Liebert, R.M. "Evaluating the Evaluators. Review of 'Sesame Street Revisited,' by T.D. Cook, H. Appleton, R. Conner, A. Shaffer, G. Tomkin, and S.J. Weber." *Journal of Communication,* 1976, *26,* (2), 165-171.
16. Huston, A.C., Wright, J.C., Wartella, E., Rice, M.L., Watkins, B.A., Campbell, T., and Potts, R. "Communicating More than Content: Formal Features of Children's Television Programs." *Journal of Communication,* 1981, *31* (3), 46.
17. Singer, J.L. "Television, Imaginative Play and Cognitive Development: Some Problems and Possibilities." Invited address to the Division of Educational Psychology, American Psychological Association, San Francisco, Calif.: 1977.
18. Tower, R.B., Singer, D.G., Singer, J.L., and Biggs, A. "Differential Effects of Television Programming on Preschoolers' Cognition, Imagination and Social Play." *American Journal of Orthopsychiatry,* 1979, *49,* (2), 265-281; Singer, D.G. "Television and Imaginative Play." *Journal of Mental Imagery,* 1978, Vol. 2 (1), 145-164.
19. Spivack, George and Shure, Myrna B. SOCIAL ADJUSTMENT OF YOUNG CHILDREN, San Francisco: Jossey-Bass, 1974.
20. Shure, Myrna B. and Spivack, George. PROBLEM-SOLVING TECHNIQUES IN CHILDREARING, San Francisco: Jossey-Bass, 1978.

6

RECOMMENDATIONS AND GUIDELINES

Each day, countless decisions are made about television. Advertisers decide what to sponsor, networks decide what to offer, producers and writers decide how to present material, and viewers everywhere decide what and how much they and their children will watch. On the assumption that detrimental as well as beneficial consequences are minimal and thus beneath notice, millions of these decision makers ignore the existing evidence and opinion. Many people assume that what is known is so inconclusive as to outweigh any concern which the knowledge might provoke. We, too, would like to see more convincing and more scientific studies. But we believe that it is neither necessary nor desirable to wait for more conclusive evidence. We know enough now to improve the quality of our decision. Aside from the fact that some of the evidence may be long in coming, we feel a sense of urgency about our society and the children we are bringing into it. While the evidence is accumulating, we continue to make decisions daily—what to air, how to present it, and what to watch.

Our special focus on children and the cumulative effects of dramatic programming, coupled with our knowledge of child development and currently available programming, lead us to make recommendations to three key groups. First, we offer guidelines for parents. Doing so, we recognize that the burden of intervention can never rest solely with parents and that, in fact, some parents, because of social and economic handicaps, lack of knowledge, or other personal conditions, may have great difficulty intervening. In particular the pressures on a single parent may present obstacles to effective monitoring of his or her child's viewing habits. These conditions highlight the role of the helping professions. We therefore direct a series of recommendations to mental health professionals and others who work with children and families. And because television has such a significant influence on society as a whole, we set forth some recommendations for public policy-makers. Finally, we make recommendations to those who make decisions in the television industry. Their creative and business decisions, after all, precede and influence all the decisions which parents and children make about television and the decisions which we make about how to help them with television.

RECOMMENDATIONS AND GUIDELINES FOR PARENTS

As the public debate about television goes on, many people, especially the parents of young children, anxiously await the outcome. Many assume that the result of the debate will be a handbook of rules and procedures whose use will reduce or eliminate the negative effects of the child's television viewing. Part of this reaction stems from the reluctance of many parents to acknowledge the ultimate fact: that they have parental responsibilities vis-á-vis the television set. Some parents appear to have withdrawn from this responsibility. They seem to assume that their children's unsupervised television watching is nothing to be concerned about, or that society or "the experts" ought to do something about the unsatisfactory programming of the networks. But the individual parental responsibility is there, and it must be exercised.

The context in which people, especially young people, watch television undoubtedly influences the impact of the material viewed. Social interaction during and after watching alters a program's significance to an individual. Its meaning is quite different from what it would have been had the person watched the same program alone. For a small child, even the passive presence of a parent or a parent figure during a frightening program can mitigate the fearful impact of the program.

The relationship of the context to the impact of viewing is an area worth much more consideration than we can give it here. In terms of our principal concern with the developing child, two points are important for parents to consider. The first consideration is the role that a balanced range of experiences plays in the promotion of healthy development.

Research findings and clinical experience suggest that lasting negative effects of television will be least for the child with a balanced range of activities and associations in a stable and structured home. In homes where educational and entertainment resources are rich, where many sources of stimulation and experience are available, and where family relations are secure, a child has learning opportunities which enable him or her to make the necessary distinctions between reality and fantasy, to learn and grow from interacting with other people, and to understand television programming in a broader context. When a child depends heavily on television for learning about the world, for fantasy material, and to gratify emotional needs, developmental delays may ensue.

The second issue a parent must consider in relation to the viewing context is the importance of his or her own participation in the child's viewing experience. There are several vital points here. One is the parent's role as a supervisor. The child needs the parent to monitor what he or she watches, to share the process of watching, to offer explanations, to provide reassurance for frightening scenes, and to help separate fact from fiction, real from imaginary. Since agreement between parents about what the child can or cannot watch is not automatic, some effort to set standards and achieve consensus on the watching patterns of the children is essential, if difficult.

The active parent helps to offset the harmful effects of problematic content and even to turn a questionable program into a growth-producing experience. Such a parent can also increase the child's benefits from positive television programming. As research has shown, the benefits of viewing positive programs are substantially greater when an adult provides interaction and support. As in other aspects of their lives together, parents help children most in dealing with experience when they guide them and interact, discuss, and share with them. The parent should be an active, influential force in the process of a child's viewing, not a passive bystander limited to making complaints about the quality of television programming.

As well as monitoring content, one or both parents should decide how much television is enough. To set no limit and leave the decision to the child encourages indiscriminate viewing. It also exposes the child to influences which may be insignificant in limited quantities but of serious consequence when viewing is frequent and prolonged. Failing to limit viewing may allow the child to lose time that should be used in diversified interpersonal and physical activities. Above all, when parents do not decide how much is enough, they abdicate an important parental responsibility—setting limits.

Last, it is worth noting that part of the problem of children's viewing habits may belong to the parents. Recent evidence suggests that the parents of many children who view excessively are themselves intemperate viewers, curiously blind to their own excess as they express concern about that of their children. In one school-initiated experiment, a number of students who were asked to turn off the television set for one week reported failure: their parents would not permit it!

Based on a study of 200 middle-class children, Singer recently reported:

> ...parents' television viewing habits were the most important predictors of children's television viewing habits. Children who spend more time watching television tend to have fathers who are heavy television watchers.[1]

A father or a working mother may see television watching as his or her due, a well-deserved relaxation after work. The parent's own wishes may thus conflict with the need to reduce the children's television-viewing time. While the parent's sharing this viewing time with the child has some positive aspects, it also risks conveying the dubious messages that television viewing is all right as long as the parent also does it, or that sharing the television viewing is the only way the child can be with the parent. Neither message is salutary, but the notion that the parent is available only when watching television is especially undesirable.

Because the parents' own viewing habits are so strongly related to those of their children and because some of the parents' viewing can and should be a shared, beneficial experience, our first recommendation to parents is that they evaluate and moderate their own television watching. Parents should temper their viewing in light of the children's needs to view less, to view what is appropriate, and, above all, to engage in real human relationships within the family.

SOME GUIDELINES

While we cannot give universal and specific rules to govern every child's everyday use of television, we can suggest some general guidelines that might be of help. Individual circumstances, attitudes, and needs must always be foremost. Thus, the guidelines are designed to aid parents in considering individual issues as they evaluate the child against some general principles.

Assess the role of television in the child's life

Spend a week or two evaluating how the child spends his or her free time, what activities he or she engages in, and how much of the time is devoted to television viewing. Consider how adequate is the range of activities, what unmet needs there might be, and what alternatives might be considered. Bear in mind the needs children have at different ages for real stimulation and activity. Consider conversation with others, books to read and to be read from, and adventures, trips, and visits. Assess the quality and quantity of play, both together with friends and alone, and opportunities for imagination, fantasy, and make-believe. Be alert to excessive imitation or mimicking of television characters and other indirect influences, such as nightmares.

Evaluate what the child is watching

Watch television with the child, and notice the kinds of programs he or she is choosing. Consider the following questions in regard to each program:

Does the program appeal to the audience for whom it was intended? Is it appropriate to the age and developmental level of the child? (Some programming designed for children may not seem appropriate for certain children. Other programs, i.e., *The Incredible Hulk,* may be very much in demand by seven-year-olds, even though these children have frequent nightmares and sleeplessness after viewing.)

Does the program present racial or ethnic groups positively? Does it show them in situations that enhance the minority group's image or the minority viewer's self-image, or does it parrot stereotypes? Does the program present sex roles and occupational adult roles fairly? (Who has the lead role? Who is the professional or leader, and who is the villain? Are the men either superheroes or incompetents? Are the women flighty and disposed to chicanery? Are teenagers portrayed with adult characteristics?)

Does the program present conflict that the child can understand, and does it present positive, nonviolent techniques for resolving conflict?

Does the program stimulate constructive activities, and does it enhance the quality of the child's play?

Does the program separate fact from fiction? Does it separate advertisements from program content?

Is the humor at the child's level, or is it adult sarcasm or ridicule, or an adult's memory of childhood amusements?

Does the pace of the program suit the readiness and developmental level of the child, allowing him or her to absorb the material presented?

Does the program present social issues that are appropriate for the child viewer, and is there something that the child can do about them? (A program discussing atomic fallout might be inappropriate, while a program on litter prevention and keeping the streets clean would be more suitable.)

Set a limit on the amount of viewing

Irrespective of program quality and appropriateness, the child may be watching too much television in a given day, week, or month. Decide how much he or she should watch in a given period.

Monitor and share in the child's viewing experiences

Continue to observe the child's viewing behavior, and be alert to any effects of the program content. Accompany the child when the program is likely to be particularly frightening or upsetting. Discuss his or her reactions, raise and answer questions, and encourage him or her to raise questions. Revise schedules and choices as needed.

Using such a checklist, parents can judge the appropriateness of particular progams for particular children. Then, as necessary, parents can assist the children, through discussion, alternative activities, and setting limits or conditions, to develop more desirable television viewing habits and to make better use of what they view. Action for Children's Television, an articulate public interest group long concerned about the detrimental effects of prolonged or inappropriate viewing, has made recommendations which they summarize in three pointed statements: "*Talk* about TV with your child! *Look* at TV with your child! *Choose* TV programs with your child!"

The process of change is not always easy and may initially result in some unhappiness or conflict. The results, however, are usually quite positive. Children adapt quickly, finding positive alternatives for the time they previously spent watching television and developing a better balance in their free-time activities. Active parental involvement in this process signifies caring, a significance that goes far beyond the immediate issue of the television dials.

The time and effort that parent and child spend discussing the programs, their content and quality, helps the child to learn to make his or her own judgment about what is best to watch. The U.S. Office of Education recently recognized the importance of having such a capacity. The Office made grants totaling $1 million to teach students how to develop critical viewing skills through the public schools. Such programs are intended to help students to "distinguish program elements, use viewing time judiciously, understand the psychological implications of advertising, distinguish fact from fiction, recognize and appeciate different views, understand the content of dramatic presentations, public affairs, news and other formats, and understand the relation between video language and the printed word."[2] It is significant that the schools are initiating such programs.

But obviously, the schools' involvement in this kind of education does not and cannot replace the parental role, particularly in teaching the young child in the home. Assessing, evaluating, and deciding about television viewing is a signal opportunity for the parent or parents to help the child develop a thoughtful, critical capacity, a capacity which will have lifelong value.

treat TV with T.L.C.

Talk about TV with your child!

- TALK ABOUT PROGRAMS THAT DELIGHT YOUR CHILD
- TALK ABOUT PROGRAMS THAT UPSET YOUR CHILD
- TALK ABOUT THE DIFFERENCES BETWEEN MAKE-BELIEVE & REAL LIFE
- TALK ABOUT WAYS TV CHARACTERS COULD SOLVE PROBLEMS WITHOUT VIOLENCE
- TALK ABOUT VIOLENCE & HOW IT HURTS
- TALK ABOUT TV FOODS THAT CAN CAUSE CAVITIES
- TALK ABOUT TV TOYS THAT MAY BREAK TOO SOON

Look at TV with your child!

- LOOK OUT FOR TV BEHAVIOR YOUR CHILD MIGHT IMITATE
- LOOK FOR TV CHARACTERS WHO CARE ABOUT OTHERS
- LOOK FOR WOMEN WHO ARE COMPETENT IN A VARIETY OF JOBS
- LOOK FOR PEOPLE FROM A VARIETY OF CULTURAL & ETHNIC GROUPS
- LOOK FOR HEALTHY SNACKS IN THE KITCHEN INSTEAD OF ON TV
- LOOK FOR IDEAS FOR WHAT TO DO WHEN YOU SWITCH OFF THE SET... READ A BOOK...DRAW A PICTURE ...PLAY A GAME

Choose TV programs with your child!

- CHOOSE THE NUMBER OF PROGRAMS YOUR CHILD CAN WATCH
- CHOOSE TO TURN THE SET OFF WHEN THE PROGRAM IS OVER
- CHOOSE TO TURN ON PUBLIC TELEVISION
- CHOOSE TO IMPROVE CHILDREN'S TV BY WRITING A LETTER TO A LOCAL STATION... TO A TELEVISION NETWORK... TO AN ADVERTISER... TO ACTION FOR CHILDREN'S TELEVISION

TENDER LOVING CARE

ACTION FOR CHILDREN'S TELEVISION
46 AUSTIN ST., NEWTONVILLE
MASS., 02160

Used with permission of Action for Children's Television, Newtonville, Mass.

RECOMMENDATIONS TO MENTAL HEALTH PROFESSIONALS

Many parents, unfortunately, cannot fulfull the responsibility to provide a range of stimulating activities for their children. For these parents and their children, television often becomes the only available stimulation. Schools, other community agencies, and those of us who counsel parents can help fill this void in cultural enrichment and also can educate children and parents about television.

Television is a major source of social and cultural information, particularly for some families. A broader public awareness of negative consequences of extended viewing, as well as the possibilities and opportunities in positive programming for all children, will enrich the public dialogue. As psychiatrists, we share with other mental health professionals the responsibility to ensure that this dialogue continues to focus on the special needs of the disadvantaged child who may view for such long hours and with so few alternatives.

Beyond this special trust, we bear the responsibility to share our observations about television and to participate in the public dialogue about its best uses, effects, and social implications. Given the economics of contemporary commerce and the stranglehold it has on programming choices, enlightened public discussion may be the best route to change. More widespread information and knowledge about television as a psychosocial issue will certainly mobilize great concern among parents, teachers, and other citizens. One result of our articulating the psychosocial issues will be a more sophisticated public. Simplistic views will be more readily challenged, such as the view that television advertising for young children is acceptable and harmless. People will come to know that children are highly susceptible to suggestion and not yet intellectually competent to assess the truth or significance of what they hear and see on television. People will realize that this limitation has implications for what ought to be considered acceptable advertising for children, if, indeed, any advertising at all is acceptable.[3]

Eventually, we must aim to make people aware that television's overall role in the socialization of children (and adults as well perhaps) is too important to take for granted. With better understanding of this remarkable phenomenon, people will be less likely to ignore the negative consequences of television or to assume that they are insignificant.

As a part of this process, special interest groups may form, working in the name of the public. Their concerns, wise or otherwise, strongly stated and vigorously advocated, may be distressing to some but will nonetheless lead to a greater sophistication about television and to more responsible decisions on the part of our society.

RECOMMENDATIONS TO PUBLIC POLICYMAKERS

While much of this report focuses on the possible detrimental effects of prolonged television viewing, we wish to add some comments about what this powerful medium of mass education might do to fulfill its potential for individual and social good.

First, however, we must consider the economics of the industry and suggest some innovations which might free the enormous pool of talent in the industry to devise more creative programming. Surely the resistance of the networks, producers, and writers to making fuller use of psychosocial knowledge must come from their struggle to obtain and hold the largest possible audience. The economics of the industry is intimately tied to the popularity of the programming. No matter how sympathetic, interested, or knowledgeable a particular writer or producer might be about the nature of children's thinking and emotional and aesthetic needs, the requirement of commercial television to attract a huge audience subordinates all other considerations. Unless or until this requirement can be reduced or contained, the likelihood seems remote that the networks, of their own volition and with their own resources, would use any other criteria for decision making, even those related to human well-being.

In the face of such an intimate relationship between economics and programming, a relationship that may be difficult or impossible to alter, we urge consideration of ways to enhance the diversity of television programming within the given situation. One step would be to increase substantially the public funding for program research, experimentation, and development. Another would be to levy a special tax on corporate television profits. A portion of the revenue could be used in place of existing public resources to support an expansion of public programming. Within this scheme, a specific percentage of money should be directed toward research into the various psychological and social effects of television. Research funds should be organized to insure the objectivity of the work and to permit significant investigation of the major priorities within the research community.[4]

RECOMMENDATIONS TO THE TELEVISION INDUSTRY

The essence of television drama in all its forms is its capacity to portray human conflict in ways which capture the interest and attention of the viewers. Drama uses a variety of elements to divert the person from a conscious awareness of his or her own life, while activating fantasies and increasing a person's accessibility to the dramatic themes.

Irrespective of the extent to which a viewer assimilates the ideas, the values, or the feelings offered on television, the quality of the presentations themselves is important, and some attention should be given to the characteristics of potentially positive programming. In our opinion, speaking particularly from the point of view of children's needs, the following recommendations will, if heeded, both improve the quality of programming and enhance its potential benefits.:

- Although it may seem banal to say so, we must recognize at the start that there is no substitute for artistry. We recommend that art and creativity be given a new priority in commercial television programming. The mind of the viewer hungers for creative quality. As others have observed, though, the insatiable maw of commercial television consumes enormous quantities of creative effort, resulting in triviality and mediocre quality. But this unfortunately does not reduce the importance of quality.
- Where television drama purports to be realistic, it should bring audiences information that is accurate, not just interesting or plausible. Medical information, for example, should be accurate. Historical docudramas should maintain a commitment to historical honesty, resisting the seductive trend to embellish the presentation with fictional rewrites which are passed off as actually having happened.
- As a consequence of the nearly ubiquitous prevalence of television viewing in the home, dramatized television characters have a special impact on young viewers. Dramatic programs in particular offer children engaging characters which can become models for children to imitate and identify with. When these characters are simplified, narrow, and constricted stereotypes of minorities, women, and others, they constitute a powerful dehumanizing influence. Viewed repeatedly, they are psychologically and socially damaging. Dramatic programs should include characters who reflect a broad range of feelings and attitudes, who illustrate complex patterns of motivation, who permit partial identification with some redeeming traits rather than flat acceptance or rejection, and who show signs of ambivalence, doubt, and other human qualities.
- Dramatic presentations should illustrate individual potential for courage, growth, self-sacrifice, commitment, and the many salient strengths of human character which can encourage, sustain, and inspire viewers. Such qualities should appear in the

context of characters and plots which are true to human reality, and not be cast in the rosy hues of a sentimental trifle.

- Television dramas ought to reconfirm the reality that human behavior is complex, that motivations are many, and that the personal and social problems to be solved have many answers. In this vein, drama should not portray human conflict as responsive to simple solutions, simple choices, snap decisions, or violence. Drama should reinforce the fact that no human being is free from anxiety, immune to conflict, or likely to live happily ever after.

- Eschewing such infantile expressions of easy living, television drama should demonstrate styles of behavior characteristic of mature human beings. Mature people work, they make an effort for what they get, they accept the differing realities of others, and they are committed to cooperating in collective approaches for solving the problems they share with others.

- Television leaders should explore the possibilities for mental health education, teaching viewers more about their emotional natures and the common interpersonal problems to which we are all heir. How such vital human knowledge could be taught by the television experiences has barely been considered.

- Because human survival depends upon perfecting our extraordinary capacity for cooperative nurturance, television drama should foster such nurturance and should discourage destructive violence. Concomitantly, treating sex as trivial or contaminating it with cruelty is inimical to fostering cooperation. However, the tasteful depiction of human love is an essential element in the wholesome presentation of the patterns of life we must try to strengthen.

A FINAL NOTE

Some people optimistically will suggest that the new telecommunication technologies are the means for improving the quality of programming. Cable, satellite, and videodisc technologies do offer the opportunity to provide audiences with more diverse fare. Since the economics of such new technologies depends less on consumer purchase of advertised products, we can hope that they will eventually contribute to a higher quality and greater diversity of television programming. Technologies, however, cannot alone face up to the programming challenges which we have described. Our challenges call for creative people, working with as much psychosocial insight as possible, to develop programs which inspire, entertain, and educate. The new technologies and their owners are and will be guided by creative people of their own, but they are under the same pressures of our commercial ethos and so are likely to draw heavily from the programming traditions of today's commercial television industry, if not directly from its programs. Because of this, the response to our challenges must come from today's industry, from ourselves, and from citizens and parents everywhere.

REFERENCES

1. Singer, D.G., Zuckerman, D.M., and Singer, J.L. Helping Elementary School Children Learn About TV. *Journal of Communication,* 30 (3): 84-93, Summer 1980.
2. Anderson, James A. The Theoretical Lineage of Critical Viewing Curricula, *Journal of Communication,* 30 (3):64-70, Summer 1980.

APPENDIX

BIBLIOGRAPHY FOR FURTHER READING

Angell, Roger. A DAY IN THE LIFE OF ROGER ANGELL. New York: Viking Press, 1970.

Baldwin, T.F., and Lewis, C. "Violence in Television: The Industry Looks at Itself." In G.A. Comstock and E.A. Rubinstein (Eds.), TELEVISION AND SOCIAL BEHAVIOR. Vol. A. Washington, D.C.: Government Printing Office, 1972, pp. 290-373.

Ball, Samuel. "Methodological Problems in Assessing the Impact of Television Programs." *Journal of Social Issues,* 1976, *32* (4), 8-17.

Bandura, A. "Social Learning Through Imitation." In M.R. Jones (Ed.), NEBRASKA SYMPOSIUM ON MOTIVATION. Lincoln, Neb.: University of Nebraska Press, 1962, pp. 211-274.

Berkowitz, L. "Violence in the Mass Media." In L. Berkowitz (Ed.), AGGRESSION: A SOCIAL PSYCHOLOGICAL ANALYSIS. New York: McGraw-Hill, 1962, pp. 229-255.

Bogatz, G.A., and Ball, S. THE SECOND YEAR OF SESAME STREET: A CONTINUING EVALUATION. Princeton, N.J.: Educational Testing Service, 1972.

Cater, Douglass, and Strickland, Stephen. TV VIOLENCE AND THE CHILD: THE EVOLUTION AND FATE OF THE SURGEON GENERAL'S REPORT. New York: Russell Sage Foundation, 1975.

Chaffee. Steven H., and Singer, Jerome L., *Television and Children,* 4 (4) /5 (1), Fall/Winter 1981/1982. (Double issue.)

Comstock, G. TELEVISION AND HUMAN BEHAVIOR: A GUIDE TO THE PERTINENT SCIENTIFIC LITERATURE. Santa Monica, Calif.: Rand Corporation, 1975, R-1746-CF, June.

Comstock, G., and Lindsey, G. TELEVISION AND HUMAN BEHAVIOR: THE RESEARCH HORIZON, FUTURE AND PRESENT. Santa Monica, Calif.: Rand Corporation, 1975, R-1748-CF, June.

Comstock, G., and Rubinstein, E.A. (Eds.) TELEVISION AND SOCIAL BEHAVIOR. VOL. 5. TECHNICAL REPORT TO THE SURGEON GENERAL'S SCIENTIFIC ADVISORY COMMITTEE ON TELEVISION AND SOCIAL BEHAVIOR. Washington, D.C.: Department of Health, Education, and Welfare, 1972.

Comstock, G., Christen, F.G., Fisher, M.L., Quarles, R.C., and

Richards, W.D. TELEVISION AND HUMAN BEHAVIOR: THE KEY STUDIES. Santa Monica, Calif.: Rand Corporation, 1975, R-1747-CF, June.

Dohrmann, Rita. "A Gender Profile of Children's Educational TV," *Journal of Communication*, 1975, *25* (4), 56-65.

Dominick, J.R. "Crime and Law Enforcement on Prime-Time Television." *Public Opinion Quarterly*, 1973, 27, 241-250.

Eron, L.D., Lefkowitz, M.M., Huesmann, L.R., and Walder, L.O. "Does Television Violence Cause Aggression?" *American Psychologist,* 1972, *27*, 253-264.

Feshbach, Seymour. "The Role of Fantasy in the Response to Television." *Journal of Social Issues,* 1976, *32* (4), 71-85.

F.C.C. Task Force on Children's TV. TV PROGRAMMING FOR CHILDREN. Washington, D.C.: Federal Communications Commission, 1979. (Five volumes).

Greenberg, B.S. "Gratifications of Television Viewing and Their Correlates for British Children." In J.G. Blumler and E. Katz (Eds.), THE USES OF MASS COMMUNICATIONS: CURRENT PERSPECTIVES ON GRATIFICATIONS RESEARCH. Beverly Hills, Calif.: Sage, 1974, 71-92.

Greenberg, B.S., and Reeves, B. "Children and the Perceived Reality of Television." *Journal of Social Issues,* 1976, *32* (4), 86-97.

Huesmann, L.R., Eron, L.D., Lefkowitz, M.M., and Walder, L.O. "Television Violence and Aggression: The Causal Effect Remains." *American Psychologist,* Does Television Violence Cause Aggression? April 1972, *27*(4), 253.

Kaplan, Robert M., and Singer, Robert. "Television Violence and Viewer Aggression: A Reexamination of the Evidence." *Journal of Social Issues,* 1976, *32* (4), 35-70.

Lange, David L., Baker, Robert K., and Ball, Sandra J. MASS MEDIA AND VIOLENCE: REPORT TO THE NATIONAL COMMISSION ON THE CAUSES AND PREVENTION OF VIOLENCE. Vol. 9. Washington, D.C.: Government Printing Office, 1969.

Lee, Barbara, "Prime Time in the Classroom" *Journal of Communication,* 1980, *30* (1), 175-180.

Lister, John. "Violence on the Screen (Report of Belson's Work)." *New England Journal of Medicine*, 1977, *297* (21), 1161-1162.

Logan, B., and Moody, K. TELEVISION AWARENESS TRAINING: A VIEWER'S GUIDE FOR FAMILY AND COMMUNITY. New York: Media Action Research Center, 1979.

Loye, David. "TV's Impact on Adults." *Psychology Today*, 1978, *11*(12), 86.

Loye, David, Gorney, R., and Steele, G. "Effects of Television: An Experimental Study." *Journal of Communication*, 1977, *27* (3), 206-216.

McCarthy, E., Langner, T.S., Gersten, J.C., Eisenberg, J.G., and Orzek, L. "Violence and Behavior Disorders." *Journal of Communication*, 1972, *25* (4), 71-85.

Milgram, S., and Shotland, R.L. TELEVISION AND ANTISOCIAL BEHAVIOR: FIELD EXPERIMENTS. New York: Academic Press, 1973.

Moody, Kathryn. "Review of 'Growing up to be Violent' " by Lefkowitz, Monroe M., Eron, Leonard D., Walder, Leopold O. and Huesmann, Rowell L. *American Journal of Orthopsychiatry*, 1978, *48*(2), 357-359.

Muson, Howard. "Teenage Violence and the Telly (Report of Belson's Work)." *Psychology Today*, 1978, *11*(10), 54-58.

Noble, Grant. "Effects of Different Forms of Filmed Aggression on Children's Constructive and Destructive Play." *Journal of Personality and Social Psychology*, 1973, *26*, 54-59.

Poulos, R.W., Rubinstein, E.A., and Liebert, R.M. "Positive Social Learning." *Journal of Communication*. 1975, *25* (4), 90-97.

Rubinstein, E.A. "Warning: The Surgeon General's Research Program May Be Dangerous to Preconceived Notions." *Journal of Social Issues*, 1976, *32* (4), 18-34.

Siegel, Alberta E. "Communicating with the Next Generation." *Journal of Communication*, 1975, *25* (4), 14-24.

Singer, Jerome L., and Singer, Dorothy G. TELEVISION, IMAGINATION, AND AGGRESSION: A STUDY OF PRESCHOOLERS. Hillsdale, N.J.: Lawrence Erlbaum Associates, 1981.

Singer, R.D., and Kaplan, R.M. "Television and Social Behavior: Introduction." *Journal of Social Issues*, 1976, *32* (4), 1-7.

Somers, Anne R. "Television and Children: Issues Involved in Corrective Action." *American Journal of Orthopsychiatry*, 1978, *48* (2), 205-213.

Thomas, M.H., Horton, R.W., Lippincott, E.C., and Drabman, R.S. "Desensitization to Portrayals of Real-Life Aggression as a Function of Exposure to Television Violence." *Journal of Personality and Social Psychology,* 1977, *35* (6), 450-458.

Winn, Marie. THE PLUG-IN DRUG. New York: Viking Press, 1977.

ACKNOWLEDGMENTS TO CONTRIBUTORS

The program of the Group for the Advancement of Psychiatry, a nonprofit, tax exempt organization, is made possible largely through the voluntary contributions and efforts of its members. For their financial assistance during the past fiscal year in helping it to fulfill its aims, GAP is grateful to the following:

Abbott Laboratories
American Charitable Foundation
Dr. and Mrs. Jeffrey Aron
Dr. and Mrs. Richard Aron
Virginia & Nathan Bederman Foundation
Ciba Pharmaceutical Company
Maurice Falk Medical Fund
Geigy Pharmaceuticals
Mrs. Carol Gold
The Gralnick Foundation
The Grove Foundation
The Holzheimer Fund
The Island Foundation
Ittleson Foundation, Inc., for Blanche F. Ittleson Consultation Program
Marion E. Kenworthy-Sarah H. Swift Foundation, Inc.
Lederle Laboratories
NcNeil Laboratories
Merck, Sharp & Dohme Laboratories
Merrell-National Laboratories
Phillips Foundation
Sandoz Pharmaceuticals
The Murray L. Silberstein Fund (Mrs. Allan H. Kalmus)
The Smith Kline Corp.
Mr. and Mrs. Herman Spertus
E.R. Squibb & Sons, Inc.
Jerome Stone Family Foundation
Tappanz Foundation
van Ameringen Foundation
Mr. S. Winn
Wyeth Laboratories